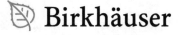

Applied and Numerical Harmonic Analysis

Series Editor
John J. Benedetto
University of Maryland
College Park, MD, USA

More information about this series at http://www.springer.com/series/4968

Árpád Bényi • Kasso A. Okoudjou

Modulation Spaces

With Applications to Pseudodifferential Operators and Nonlinear Schrödinger Equations

 Birkhäuser

Árpád Bényi
Department of Mathematics
Western Washington University
Bellingham, WA, USA

Kasso A. Okoudjou
Department of Mathematics
University of Maryland
College Park, MD, USA

ISSN 2296-5009   ISSN 2296-5017   (electronic)
Applied and Numerical Harmonic Analysis
ISBN 978-1-0716-0614-8   ISBN 978-1-0716-0332-1   (eBook)
https://doi.org/10.1007/978-1-0716-0332-1

Mathematics Subject Classification: 42-02, 42B15, 42B35, 42B37, 47G30, 35Q55, 46E30, 46E35, 35-XX, 46-XX

This book is published under the imprint Birkhäuser, www.birkhauser-science.com by the registered company Springer Science+Business Media, LLC
The registered company address is: 1 New York Plaza, New York, NY 10004, U.S.A.

Pentru Mona, Alex şi Basti, cu multă dragoste-Á.B.
Á Rouky, Shadeh, Shola, et Femi, avec amour et affection-K.A.O.

ANHA Series Preface

The *Applied and Numerical Harmonic Analysis (ANHA)* book series aims to provide the engineering, mathematical, and scientific communities with significant developments in harmonic analysis, ranging from abstract harmonic analysis to basic applications. The title of the series reflects the importance of applications and numerical implementation, but richness and relevance of applications and implementation depend fundamentally on the structure and depth of theoretical underpinnings. Thus, from our point of view, the interleaving of theory and applications and their creative symbiotic evolution is axiomatic.

Harmonic analysis is a wellspring of ideas and applicability that has flourished, developed, and deepened over time within many disciplines and by means of creative cross-fertilization with diverse areas. The intricate and fundamental relationship between harmonic analysis and fields such as signal processing, partial differential equations (PDEs), and image processing is reflected in our state-of-the-art *ANHA* series.

Our vision of modern harmonic analysis includes mathematical areas such as wavelet theory, Banach algebras, classical Fourier analysis, time-frequency analysis, and fractal geometry, as well as the diverse topics that impinge on them.

For example, wavelet theory can be considered an appropriate tool to deal with some basic problems in digital signal processing, speech and image processing, geophysics, pattern recognition, biomedical engineering, and turbulence. These areas implement the latest technology from sampling methods on surfaces to fast algorithms and computer vision methods. The underlying mathematics of wavelet theory depends not only on classical Fourier analysis, but also on ideas from abstract harmonic analysis, including von Neumann algebras and the affine group. This leads to a study of the Heisenberg group and its relationship to Gabor systems, and of the metaplectic group for a meaningful interaction of signal decomposition methods. The unifying influence of wavelet theory in the aforementioned topics illustrates the justification for providing a means for centralizing and disseminating information from the broader, but still focused, area of harmonic analysis. This will be a key role of *ANHA*. We intend to publish with the scope and interaction that such a host of issues demands.

Along with our commitment to publish mathematically significant works at the frontiers of harmonic analysis, we have a comparably strong commitment to publish major advances in the following applicable topics in which harmonic analysis plays a substantial role:

Antenna theory	*Prediction theory*
Biomedical signal processing	*Radar applications*
Digital signal processing	*Sampling theory*
Fast algorithms	*Spectral estimation*
Gabor theory and applications	*Speech processing*
Image processing	*Time-frequency and*
Numerical partial differential equations	*Time-scale analysis*
	Wavelet theory

The above point of view for the *ANHA* book series is inspired by the history of Fourier analysis itself, whose tentacles reach into so many fields.

In the last two centuries Fourier analysis has had a major impact on the development of mathematics, on the understanding of many engineering and scientific phenomena, and on the solution of some of the most important problems in mathematics and the sciences. Historically, Fourier series were developed in the analysis of some of the classical PDEs of mathematical physics; these series were used to solve such equations. In order to understand Fourier series and the kinds of solutions they could represent, some of the most basic notions of analysis were defined, e.g., the concept of "function." Since the coefficients of Fourier series are integrals, it is no surprise that Riemann integrals were conceived to deal with uniqueness properties of trigonometric series. Cantor's set theory was also developed because of such uniqueness questions.

A basic problem in Fourier analysis is to show how complicated phenomena, such as sound waves, can be described in terms of elementary harmonics. There are two aspects of this problem: first, to find, or even define properly, the harmonics or spectrum of a given phenomenon, e.g., the spectroscopy problem in optics; second, to determine which phenomena can be constructed from given classes of harmonics, as done, for example, by the mechanical synthesizers in tidal analysis.

Fourier analysis is also the natural setting for many other problems in engineering, mathematics, and the sciences. For example, Wiener's Tauberian theorem in Fourier analysis not only characterizes the behavior of the prime numbers, but also provides the proper notion of spectrum for phenomena such as white light; this latter process leads to the Fourier analysis associated with correlation functions in filtering and prediction problems, and these problems, in turn, deal naturally with Hardy spaces in the theory of complex variables.

Nowadays, some of the theory of PDEs has given way to the study of Fourier integral operators. Problems in antenna theory are studied in terms of unimodular trigonometric polynomials. Applications of Fourier analysis abound in signal processing, whether with the fast Fourier transform (FFT), or filter design, or the adaptive modeling inherent in time-frequency-scale methods such as wavelet theory. The coherent states of mathematical physics are translated and modulated Fourier transforms, and these are used, in conjunction with the uncertainty principle, for dealing with signal reconstruction in communications theory. We are back to the raison d'être of the *ANHA* series!

University of Maryland   John J. Benedetto
College Park, MD, USA   Series Editor

Preface

There are almost two decades since the publication of the by now classical text of Gröchenig [125] on the fundamentals of time-frequency analysis. During this period, a considerable amount of research has been dedicated to the study of "good" function spaces that treat time and frequency simultaneously as well as those operators acting continuously on them. In some sense, this book begins where [125] ends. Its main goal is to give a flavor of *some* of the latest developments around the topic of modulation spaces defined on the d-dimensional Euclidean space, by establishing their basic properties, equivalence of various definitions that have appeared in the literature, and some of their applications to pseudodifferential operators and partial differential equations. In fact, these applications are the raison d'être of this monograph. Indeed, our journey to the subject started from the investigation of the boundedness properties on the modulation spaces of certain multilinear pseudodifferential operators that are known to be unbounded on most classical function spaces. In the process and through the years, we came across a variety of applications of modulation spaces to not only pseudodifferential operators, but also in nonlinear partial differential equations. Therefore, many of the results discussed in this text relate in one form or another to some of the work done by the authors on the topics presented justifying further that its scope is not to be comprehensive, as indeed it cannot be. The notes at the end of most chapters contain connections with other interesting related topics to the ones touched upon within the text and some references useful in exploring such venues.

Because this monograph is shaped by the aforementioned applications, we expect that it will be useful for researchers in both time-frequency analysis and partial differential equations. Furthermore, our hope is that this book will serve as a standard reference to graduate students that are interested in the study of modulation spaces, but seasoned researchers should also find relevant the account of some of the main developments in this area of time-frequency analysis and the overarching sense of unity between several of these topics. The titles of each chapter and the sections within should be self-explanatory to their purpose.

Parts of this book have been used in the first author's teaching of one quarter graduate (second year masters) level topics in analysis courses, but the intention of these pages was to constitute a research monograph, rather than a textbook. We tried to write a text that is as self-contained as possible. However, in certain instances we chose brevity and stated certain results without giving a proof or only sketched the argument, and preferred to refer for a full proof to the original source. We hope that the interested reader will ultimately find this approach of outlining the main ideas underlying a deeper statement advantageous.

In terms of organization, we note that Chap. 1 contains a number of foundational results and facts in functional, harmonic, and real analysis. Chapters 2 and 3 give an exposition of various equivalent definitions of the unweighted modulation spaces. The natural connection between the modulation spaces and the theory of (linear and multilinear) pseudodifferential operators is the content of Chap. 4, which can be thought of as one of our main motivations on embarking in the project of writing this monograph. Chapter 5 introduces the weighted modulation spaces and can be viewed as a very interesting application of the theory of linear pseudodifferential operators developed in Chap. 4. Chapter 6 deals with the embeddings between the modulation spaces and other function spaces and offers some natural conditions on functions or distributions to belong to the former. Finally, Chap. 7 is a preview of the untapped and potential roles of the modulation spaces in the analysis of dispersive partial differential equations and focuses solely on the nonlinear Schrödinger equations. On a structural level, this last chapter can be seen as an alluring application of the materials previously developed in the text.

The authors met in February 2002 while they were still graduate students at the 5th New Mexico Analysis Seminar, held at New Mexico State University, Las Cruces. It was at this meeting that Kasso suggested to Árpád that some of his work would be better suited to the realm of modulation spaces, a suggestion that led to several joint collaborations. This text then is fittingly a celebration of the authors long-time friendship inside and outside of mathematics.

According to Lieb and Loss [178, Preface to the First Edition], a book is one of the things "that are easy to start but very difficult to finish." We certainly share this feeling. This book has been a long way coming and its completion would not have been possible without the love and support of our families. These pages are dedicated to them.

The first author gratefully acknowledges the support of the Simons Foundation (Grant No. 246024 to Árpád Bényi) and his institution, Western Washington University, for the sabbatical leaves which benefited the writing of this text. The second author was partially supported by a grant from the Simons Foundation (Grant No. 319197 to Kasso Okoudjou), the National Science Foundation under Grant No. DMS-1814253, and an MLK visiting professorship at MIT.

Bellingham, WA, USA   Árpád Bényi
College Park, MD, USA   Kasso A. Okoudjou
April 2019

Acknowledgments

We would like to thank John Benedetto for suggesting us to write this text, to Hans Feichtinger for his encouraging words at the onset of the project, and to Chris Heil and Karlheinz Gröchenig for their guidance when we were just getting started on our respective journeys in the realm of time-frequency analysis.

I am grateful to Rodolfo H. Torres, Tadahiro Oh, and Branko Ćurgus for their longtime friendship and for their continued inspiration in mathematics.-Á.B.

I am indebted to Chris Heil for introducing me to this field and to John Benedetto and Karlheinz Gröchenig for their constant support and encouragement.-K.A.O.

Contents

Chapter 1
Notions of Real, Functional, and Fourier Analysis

The definitions and results of real, functional, and harmonic analysis presented here should be viewed as a minimal prerequisite. The reader interested in learning more about any of these subjects is referred to the books of Benedetto [4], Benedetto and Czaja [5], Folland [109], Rudin [200], Grafakos [119, 120], and Gröchenig [125].

1.1 Real Analysis

The underlying space we will work with is the Euclidean space $\mathbb{R}^d$. For $x = (x_1, x_2, \ldots, x_d) \in \mathbb{R}^d$, $dx = dx_1 dx_2 \cdots dx_d$ denotes the Lebesgue measure, while $\int_{\mathbb{R}^d} f(x)\, dx$ is the usual Lebesgue integral of a function $f : \mathbb{R}^d \to \mathbb{C}$. The characteristic function of a measurable set $\mathbb{S} \subset \mathbb{R}^d$ is denoted by $\chi_\mathbb{S}$, while the measure of such a set is given by $|\mathbb{S}| = \int_{\mathbb{R}^d} \chi_\mathbb{S}\, dx$. The notation $\mathbb{B}(x, r) = \{y \in \mathbb{R}^d : |y - x|_d < r\}$ will stand for the open ball in $\mathbb{R}^d$ centered at $x \in \mathbb{R}^d$ and of radius $r > 0$, where we wrote $|x|_d = (x_1^2 + x_2^2 + \cdots + x_d^2)^{1/2}$ for the Euclidean norm on $\mathbb{R}^d$. For ease of notation, we will suppress the index in $|\cdot|_d$ and simply write $|\cdot|$, with the meaning being unambiguous from the context; for example, if $x \in \mathbb{R}$, then $|x|$ is the absolute value of x, if $z = x + iy \in \mathbb{C}$, then $|z| = |(x, y)| = (x^2 + y^2)^{1/2}$ is the modulus of z, etc. The closed ball centered at x and of radius r will be denoted by $\overline{\mathbb{B}(x, r)}$.

1.1.1 L^p Spaces

Definition 1.1 For $0 < p \leq \infty$, $L^p(\mathbb{R}^d)$ (or simply L^p, when there is no ambiguity) denotes the set of all (Lebesgue) measurable functions $f : \mathbb{R}^d \to \mathbb{C}$ such that the L^p quasi-norm $\|f\|_{L^p}$ is finite, where we denoted

© Springer Science+Business Media, LLC, part of Springer Nature 2020
Á. Bényi, K. A. Okoudjou, *Modulation Spaces*, Applied and Numerical Harmonic Analysis, https://doi.org/10.1007/978-1-0716-0332-1_1

$$\|f\|_{L^p} = \left(\int_{\mathbb{R}^d} |f(x)|^p \, dx\right)^{1/p}, \text{ if } 0 < p < \infty;$$

$$\|f\|_{L^\infty} = \text{esssup}_{x \in \mathbb{R}^d} |f(x)| := \inf\{m > 0 : |\{x \in \mathbb{R}^d : |f(x)| > m\}| = 0\}, \text{ if } p = \infty.$$

Similarly, for a measurable set $\mathbb{S} \subset \mathbb{R}^d$, we say that $f \in L^p(\mathbb{S})$ if and only if $f \chi_\mathbb{S} \in L^p(\mathbb{R}^d)$.

For two functions $f, g \in L^p$, the equality $f = g$ means that f and g are equal almost everywhere. If $1 \leq p \leq \infty$, the L^p spaces are Banach spaces. For $0 < p < 1$ they are only quasi-Banach spaces, that is, they are complete but the triangle inequality is now replaced by the quasi-triangle inequality

$$\|f + g\|_{L^p} \leq 2^{-p'} \left(\|f\|_{L^p} + \|g\|_{L^p}\right).$$

We use the notation $p' = p/(p - 1)$ for any $p \in (0, \infty)$, $p \neq 1$. We also set $1' = \infty$ and $\infty' = 1$. Note that, in particular, $(p')' = p$ for all $p \in (0, \infty]$. In general, for a measurable set $\mathbb{S} \subset \mathbb{R}^d$, we do not have any embedding property available between the spaces $L^p(\mathbb{S})$ and $L^q(\mathbb{S})$ for $p \neq q$. However, if we further assume that $0 < p < q < \infty$ and $|\mathbb{S}| < \infty$, then $L^\infty(\mathbb{S}) \subset L^q(\mathbb{S}) \subset L^p(\mathbb{S})$ and

$$\|f\|_{L^p(\mathbb{S})} \leq |\mathbb{S}|^{\frac{1}{p} - \frac{1}{q}} \|f\|_{L^q(\mathbb{S})}.$$

This fact is an immediate consequence of the following result.

Theorem 1.2 (Hölder's Inequality) *Given two measurable functions f, g on $\mathbb{R}^d$, and $0 < p, q, r \leq \infty$ such that $1/r = 1/p + 1/q$, we have*

$$\|fg\|_{L^r} \leq \|f\|_{L^p} \|g\|_{L^q}. \tag{1.1}$$

Hölder's inequality can be extended to any finite product of k functions, where $k \in \mathbb{N}, k \geq 2$:

If $f_j \in L^{p_j}, 1 \leq j \leq k$, and $1/p = \sum_{j=1}^k 1/p_j$ with $0 < p, p_1, \ldots, p_k \leq \infty$, then $\prod_{j=1}^k f_j \in L^p$ and

$$\left\|\prod_{j=1}^k f_j\right\|_{L^p} \leq \prod_{j=1}^k \|f_j\|_{L^{p_j}}. \tag{1.2}$$

An important particular case of inequality (1.1) is obtained by letting $q = p'$ and $r = 1$:

$$\|fg\|_{L^1} \leq \|f\|_{L^p} \|g\|_{L^{p'}}. \tag{1.3}$$

Letting $p = 2$ in (1.3) we obtain the so-called *Cauchy-Bunyakowsky-Schwarz inequality.* Moreover, for all $1 \leq p < \infty$, we have the equality

$$\|f\|_{L^p} = \sup_{\|g\|_{L^{p'}} = 1} \left|\int_{\mathbb{R}^d} f(x)\overline{g(x)} \, dx\right|,$$

which establishes that the spaces L^p and $L^{p'}$ are dual to each other; we will define precisely the notion of *dual space* in the next section, see Definition 1.22. The space L^2 plays a special role in analysis; because $2' = 2$, this space is dual to itself and it is the only Hilbert space in the family of Lebesgue spaces $\{L^p : 0 < p \leq \infty\}$, endowed with the inner product

$$\langle f, g \rangle_{L^2} = \int_{\mathbb{R}^d} f(x)\overline{g(x)}\,dx.$$

A tool that will be used repeatedly throughout the subsequent chapters is contained in the following statement.

Theorem 1.3 (Minkowski's Integral Inequality) *For all* $p \in [1, \infty)$ *and all measurable functions* $F(x, y)$ *on the product space* $\mathbb{R}^d \times \mathbb{R}^n$ *we have*

$$\left\| \left\| F(x, y) \right\|_{L_x^1} \right\|_{L_y^p} \leq \left\| \left\| F(x, y) \right\|_{L_y^p} \right\|_{L_x^1}. \tag{1.4}$$

Written out explicitly, inequality (1.4) is:

$$\left(\int_{\mathbb{R}^n} \left(\int_{\mathbb{R}^d} |F(x, y)|\,dx \right)^p dy \right)^{1/p} \leq \int_{\mathbb{R}^d} \left(\int_{\mathbb{R}^n} |F(x, y)|^p\,dy \right)^{1/p} dx.$$

When $0 < p < 1$, the inequality in (1.4) is reversed.

An inequality of a somewhat similar flavor is known under the name of *Fatou's lemma*.

Theorem 1.4 (Fatou) *Let* $0 < p < \infty$ *and* $(f_j)_{j \in \mathbb{N}}$ *a sequence of measurable functions on* $\mathbb{R}^d$. *Then, we have*

$$\left\| \liminf_{j \to \infty} f_j \right\|_{L^p} \leq \liminf_{j \to \infty} \|f_j\|_{L^p}.$$

In general, given a sequence (f_j) of measurable functions on $\mathbb{R}^d$, the almost everywhere convergence of the sequence to some function f does not immediately imply the convergence of $\int_{\mathbb{R}^d} f_j(x)\,dx$ to $\int_{\mathbb{R}^d} f(x)\,dx$ as the cautionary example (with $d = 1$) $f_j = \chi_{[j, j+1)}$ and $f = 0$ shows. Naturally, we can ask when does almost everywhere convergence guarantee convergence in L^p. We have the following.

Theorem 1.5 (Lebesgue's Dominated Convergence) *If* $f_j \in L^p$, $f_j \to f$ *a.e. as* $j \to \infty$, *and there exists* $g \in L^p$, $g \geq 0$, *such that* $|f_j| \leq g$ *a.e. for all* $j \in \mathbb{N}$, *then* $f \in L^p$ *and* $\|f_j - f\|_{L^p} \to 0$ *as* $j \to \infty$.

The statements of Minkowski's inequality and Fatou's lemma imply some change in the order in which each operation (integration or limit) is performed. Of course, in general, it is not possible to simply change the order of integration in a double integral. However, there are results that give sufficient conditions for when this interchange is allowed.

Theorem 1.6 (Tonelli) *If $F(x, y) \geq 0$ on $\mathbb{R}^d \times \mathbb{R}^n$, then*

$$\int_{\mathbb{R}^n} \left(\int_{\mathbb{R}^d} F(x, y)\, dx \right) dy = \int_{\mathbb{R}^d} \left(\int_{\mathbb{R}^n} F(x, y)\, dy \right) dx. \tag{1.5}$$

In fact, in this case, both iterated integrals equal the double integral

$$\int \int_{\mathbb{R}^d \times \mathbb{R}^n} F(x, y)\, dx dy.$$

Another sufficient condition for the change in integration to work is given by *Fubini's theorem.*

Theorem 1.7 (Fubini) *Assume $F(x, y) \in L^1(\mathbb{R}^d \times \mathbb{R}^n)$. Then, for almost every $y \in \mathbb{R}^n$, the section $x \mapsto F(x, y)$ is in $L^1(\mathbb{R}^d)$ and for almost every $x \in \mathbb{R}^d$, the section $y \mapsto F(x, y)$ is in $L^1(\mathbb{R}^n)$. Moreover, the function $x \mapsto \int_{\mathbb{R}^n} F(x, y)\, dy$ is in $L^1(\mathbb{R}^d)$, the function $y \mapsto \int_{\mathbb{R}^d} F(x, y)\, dx$ is in $L^1(\mathbb{R}^n)$ and equality (1.5) holds.*

The definition of modulation spaces depends on the notion of mixed Lebesgue spaces, which we introduce next.

Definition 1.8 For $0 < p, q \leq \infty$, $L^p L^q(\mathbb{R}^{2d})$ (or $L^p L^q$, or simply $L^{p,q}$ when there is no ambiguity) denotes the set of all (Lebesgue) measurable functions $F : \mathbb{R}^{2d} \to \mathbb{C}$ such that the $L^p L^q$ quasi-norm $\|F\|_{L^p L^q}$ is finite, where

$$\|F\|_{L^p L^q} = \left(\int_{\mathbb{R}^d} \left(\int_{\mathbb{R}^d} |F(x, y)|^p\, dx \right)^{q/p} dy \right)^{1/q}, \text{ if } 0 < p, q < \infty;$$

$$\|F\|_{L^\infty L^q} = \left(\int_{\mathbb{R}^d} \left(\operatorname{esssup}_{x \in \mathbb{R}^d} |F(x, y)|^q \right) dy \right)^{1/q}, \text{ if } 0 < q < \infty;$$

$$\|F\|_{L^p L^\infty} = \operatorname{esssup}_{y \in \mathbb{R}^d} \left(\int_{\mathbb{R}^d} |F(x, y)|^p\, dx \right)^{1/p}, \text{ if } 0 < p < \infty.$$

As for Lebesgue spaces, the mixed Lebesgue spaces are quasi-Banach spaces (Banach spaces, if $1 \leq p, q \leq \infty$) and their duality follows a familiar pattern. For $1 \leq p, q < \infty$, one can show that $L^{p', q'}$ is the dual of $L^{p,q}$, and this follows now from *Hölder's inequality for mixed Lebesgue spaces*, in particular, the counterpart to inequality (1.3).

Proposition 1.9 *If $F \in L^{p,q}$ and $G \in L^{p', q'}$, then $FG \in L^1$ and*

$$\|FG\|_{L^1} \leq \|F\|_{L^{p,q}} \|G\|_{L^{p', q'}}. \tag{1.6}$$

While our previous discussion was concerned strictly with Lebesgue spaces on $\mathbb{R}^d$ endowed with the usual Lebesgue measure, all the previous results apply to $L^p(\mathbb{X}, \mu)$ spaces, where $\mathbb{X}$ is a measure space and μ is a positive measure on it. In the rare situations in which we will encounter this general framework, we shall point out the necessary adjustments (or details) needed.

We end this subsection by briefly recalling the notion of *local integrability*. Similar to Definition 1.1, for all $0 < p < \infty$ we can define the space $L_{\text{loc}}^p(\mathbb{R}^d)$ (or simply L_{loc}^p) as the set of all measurable functions f on $\mathbb{R}^d$ such that for any compact set $\mathbb{K} \subset \mathbb{R}^d$ we have $\int_{\mathbb{K}} |f(x)|^p \, dx < \infty$. Functions in L_{loc}^1 are called locally integrable. These local versions of the Lebesgue spaces are nested: for $0 < p < q < \infty$, we have $L_{\text{loc}}^q \subset L_{\text{loc}}^p$. It is also easy to see that

$$\bigcup_{p \geq 1} L^p \subset L_{\text{loc}}^1.$$

For $0 < p < 1$, the inclusion $L^p(\mathbb{R}^d) \subset L_{\text{loc}}^1(\mathbb{R}^d)$ fails in general, as the example $f(x) = |x|^{-d-\epsilon} \chi_{\overline{\mathbb{B}(0,1)}}(x)$, $\epsilon > 0$, shows.

1.1.2 Convolution

We dedicate this subsection to a brief review of the notion of convolution of two functions. This notion can be defined in general on topological groups (that are not necessarily abelian), but we restrict again our attention to $\mathbb{R}^d$.

Definition 1.10 For $f, g \in L^1(\mathbb{R}^d)$, the convolution $f * g$ is defined by

$$(f * g)(x) = \int_{\mathbb{R}^d} f(y)g(x - y) \, dy.$$

$(L^1, *)$ is a Banach algebra. In particular, $*$ is a commutative and associative operation. Two immediate consequences of the definition are that the translate of the convolution of two functions is the same as the convolution of one of the functions translated and the other one left intact, while the support of the convolution of two functions is a subset of the set addition of the supports of these functions. Also, it is easy to see that the convolution of two integrable functions is an integrable function as well and $\|f * g\|_{L^1} \leq \|f\|_{L^1} \|g\|_{L^1}$. In fact, this inequality can be generalized into what is commonly known as *Young's inequality*.

Theorem 1.11 (Young's Inequality) *If $p, q, r \in [1, \infty]$ satisfy $1/p + 1/q = 1 + 1/r$, then for all $f \in L^p$ and $g \in L^q$ we have*

$$\|f * g\|_{L^r} \leq \|f\|_{L^p} \|g\|_{L^q}. \tag{1.7}$$

Mathematical induction allows us to extend Young's inequality to an arbitrary convolution of k functions, where $k \in \mathbb{N}, k \geq 2$.

If $f_j \in L^{p_j}, 1 \leq j \leq k$, and $\sum_{j=1}^{k} 1/p_j = k - 1 + 1/p$ with $1 \leq p, p_1, \ldots, p_k \leq \infty$, then $f_1 * \cdots * f_k \in L^p$ and

$$\left\| f_1 * \cdots * f_k \right\|_{L^p} \leq \prod_{j=1}^{k} \left\| f_j \right\|_{L^{p_j}}. \tag{1.8}$$

The proof of (1.7) is a nice application of the Hölder inequality (1.1) and Fubini's Theorem 1.7. We present it here for the benefit of the reader.

Proof Note first that if, say, $q = \infty$, then necessarily $p = 1, r = \infty$ and the inequality is trivial to prove. Thus, let us assume $q < \infty$. Now, the condition on the indices can be rewritten as $1/p = 1/q' + 1/r$ or $1/q = 1/p' + 1/r$ or $1/q' + 1/r + 1/p = 1$. In particular, $(p/r)' = p/q'$ and $(q/r)' = q/p'$. Next, we split the product $|f(y)||g(x - y)|$ as a product of three terms

$$|f(y)|^{p/q'} \left(|f(y)|^{p/r} |g(x - y)|^{q/r} \right) |g(x - y)|^{q/p'}$$

and apply Hölder's inequality (1.2) with exponents q', r, and p' to this product. We get

$$|(f * g)(x)| \leq \|f\|_{L^p}^{p/q'} \left(\int_{\mathbb{R}^d} |f(y)|^p |g(x - y)|^r \, dy \right)^{1/r} \left(\int_{\mathbb{R}^d} |g(x - y)|^q \, dy \right)^{1/p'}$$

$$= \|f\|_{L^p}^{p/q'} \|g\|_{L^q}^{q/p'} \left(\int_{\mathbb{R}^d} |f(y)|^p |g(x - y)|^r \, dy \right)^{1/r}.$$

Now, we are interested in the L^r norm of $f * g$, that is, $\left(\int_{\mathbb{R}^d} |(f * g)(x)|^r \right)^{1/r}$. The previous inequality and Fubini's Theorem 1.7 give

$$\|f * g\|_{L^r} \leq \|f\|_{L^p}^{p/q'} \|g\|_{L^q}^{q/p'} \left(\int_{\mathbb{R}^d} \int_{\mathbb{R}^d} |f(y)|^p |g(x - y)|^r \, dx dy \right)^{1/r}$$

$$= \|f\|_{L^p}^{p/q'} \|g\|_{L^q}^{q/p'} \|f\|_{L^p}^{p/r} \|g\|_{L^q}^{q/r}$$

$$= \|f\|_{L^p} \|g\|_{L^q}.$$

In the particular case $r = \infty$, that is, $f \in L^p$ and $g \in L^{p'}$, we further have that $f * g$ is uniformly continuous; moreover, if $1 < p < \infty$, then $(f * g)(x) \to 0$ as $|x| \to \infty$.   □

1.1.3 ℓ^p and $\ell^{p,q}$ Spaces

On the space $\mathbb{Z}^d$ we have the counting measure $\nu = \sum_{k \in \mathbb{Z}^d} \delta_k$, where δ_k denotes the Dirac measure concentrated at k. Integration with respect to ν gives summation over the lattice $\mathbb{Z}^d$:

$$\int_{\mathbb{R}^d} f \, d\nu = \sum_{k \in \mathbb{Z}^d} f(k).$$

Definition 1.12 For $0 < p \leq \infty$, the space $L^p(\mathbb{Z}^d, \nu)$ is denoted by $\ell^p(\mathbb{Z}^d)$ (or simply ℓ^p when there is no ambiguity).

In analogy with the mixed Lebesgue spaces $L^{p,q}(\mathbb{R}^{2d})$, we can then define the mixed spaces $\ell^p \ell^q(\mathbb{Z}^{2d})$ (or $\ell^p \ell^q$, or simply $\ell^{p,q}$). They consist now of functions $\mathbf{s} : \mathbb{Z}^{2d} \to \mathbb{C}$ such that the $\ell^p \ell^q$ quasi-norm $\|\mathbf{s}\|_{\ell^p \ell^q}$ is finite. Of course, $\mathbf{s}$ is nothing else but a doubly indexed sequence $\mathbf{s} = (s_{kl})_{k,l \in \mathbb{Z}^d}$. Thus, for $0 < p, q < \infty$, we have

$$\|\mathbf{s}\|_{\ell^p \ell^q} = \left(\sum_{k \in \mathbb{Z}^d} \left(\sum_{l \in \mathbb{Z}^d} |s_{kl}|^p \right)^{q/p} \right)^{1/q}.$$

Moreover, one can also define the mixed Lebesgue spaces $L^p \ell^q$ and $\ell^q L^p$. For example, in the first case, we are talking about sequences of functions $\mathbf{f} = (f_k)_{k \in \mathbb{Z}^d}$ such that the quasi-norm $\|\mathbf{f}\|_{L^p \ell^q}$ is finite, where

$$\|\mathbf{f}\|_{L^p \ell^q} = \|(\|f_k\|_{L^p})_{k \in \mathbb{Z}^d}\|_{\ell^q}$$

$$= \left(\sum_{k \in \mathbb{Z}^d} \|f_k\|_{L^p}^q \right)^{1/q} = \left(\sum_{k \in \mathbb{Z}^d} \left(\int_{\mathbb{R}^d} |f_k(x)|^p dx \right)^{q/p} \right)^{1/q}.$$

When $p = \infty$ or $q = \infty$ the essential supremum is used in all the expressions of the quasi-norms above. As expected, all the quasi-norms discussed are also complete, thus turning all the flavors of mixed Lebesgue spaces into quasi-Banach spaces (Banach spaces, if $1 \leq p, q \leq \infty$).

In view of the previous two sections, and remembering that most of our previous discussion transfers verbatim to L^p spaces of arbitrary measure spaces, we also obtain a plethora of properties of ℓ^p spaces. For example, if in Fubini's theorem one or both of the measures are replaced by the counting measure ν, we obtain some special cases in which it is permissible to interchange the operations of integration and summation.

Proposition 1.13

(a) *If $f_j \in L^1(\mathbb{R}^d)$ are such that $\sum_{k \in \mathbb{Z}^d} \|f_j\|_{L^1}$ is convergent, then*

$$\int_{\mathbb{R}^d} \left(\sum_{j \in \mathbb{Z}^d} f_j(x) \right) dx = \sum_{j \in \mathbb{Z}^d} \int_{\mathbb{R}^d} f_j(x) \, dx.$$

(b) *If* $\mathbf{s} = (s_{kl}) \in \ell^{1,1}(\mathbb{Z}^{2d})$, *then*

$$\sum_{k \in \mathbb{Z}^d} \left(\sum_{l \in \mathbb{Z}^d} s_{kl} \right) = \sum_{l \in \mathbb{Z}^d} \left(\sum_{k \in \mathbb{Z}^d} s_{kl} \right) = \sum_{(k,l) \in \mathbb{Z}^{2d}} s_{kl}.$$

The following lemma states some elementary inequalities for finitely supported sequences. Despite its simplicity, it has some important consequences which we point out immediately after its statement; moreover, and more importantly, we will make use of this result in Chap. 4, specifically in our discussion of modulation spaces with indices below 1.

Lemma 1.14 *Let* $\mathbf{a} = (a_k)_{k \in \mathbb{N}}$ *be a sequence of complex numbers,* $n \geq 1$ *a fixed integer, and* $p \geq 0$.

(a) *If* $1 \leq p < \infty$, *then*

$$\left(\sum_{k=1}^{n} |a_k| \right)^p \leq n^{p-1} \sum_{k=1}^{n} |a_k|^p \quad and \quad \sum_{k=1}^{n} |a_k|^p \leq \left(\sum_{k=1}^{n} |a_k| \right)^p.$$

(b) *If* $0 \leq p < 1$, *then*

$$\sum_{k=1}^{n} |a_k|^p \leq n^{-p+1} \left(\sum_{k=1}^{n} |a_k| \right)^p \quad and \quad \left(\sum_{k=1}^{n} |a_k| \right)^p \leq \sum_{k=1}^{n} |a_k|^p.$$

An application of Lemma 1.14 yields the following *Minkowski inequalities*; the reader should compare the following statement to the comments immediately following Definition 1.1.

Theorem 1.15 *Let* $(f_k)_{k \in \mathbb{N}}$ *be a sequence of* L^p *functions.*

(i) *If* $1 \leq p \leq \infty$, *then*

$$\left\| \sum_{k=1}^{\infty} f_k \right\|_{L^p} \leq \sum_{k=1}^{\infty} \| f_k \|_{L^p}.$$

(ii) *If* $0 < p < 1$, *then*

$$\sum_{k=1}^{\infty} \| f_k \|_{L^p} \leq \left\| \sum_{k=1}^{\infty} |f_k| \right\|_{L^p}.$$

An important property of the ℓ^p spaces is their *nestedness*.

Proposition 1.16 *If $0 < p < q \leq \infty$, then $\ell^p(\mathbb{Z}^d) \subset \ell^q(\mathbb{Z}^d)$.*

Proof We sketch the argument only for $d = 1$. Let $\mathbf{a} = (a_k)_{k\in\mathbb{Z}}$ be such that $\mathbf{a} \in \ell^p$. Fix $N \in \mathbb{N}$. Then, $\sum_{k=-N}^{N} |a_k|^p \leq \|\mathbf{a}\|_{\ell^p}^p < \infty$. Now, using Lemma 1.14 and the fact that $q/p > 1$, we can write

$$\sum_{k=-N}^{N} |a_k|^q \leq \left(\sum_{k=-N}^{N} |a_k|^p \right)^{q/p} \leq \|\mathbf{a}\|_{\ell^p}^q.$$

Thus, letting $N \to \infty$, we get $\|\mathbf{a}\|_{\ell^q} \leq \|\mathbf{a}\|_{\ell^p}$. $\qquad\qquad\square$

The inclusion between ℓ^p spaces pointed out in Proposition 1.16 is strict. For example, letting $\mathbf{a} = (1/k)_{k\in\mathbb{Z}}$, we have $\mathbf{a} \in \ell^2 \setminus \ell^1$. Also, an immediate consequence of Proposition 1.16, is the *monotonicity property of $\ell^p\ell^q$ spaces*; we will make use of this simple property in Chap. 2, when dealing with the inclusion relations between modulation spaces.

Proposition 1.17 *If $0 < p_1 \leq p_2 \leq \infty$ and $0 < q_1 \leq q_2 \leq \infty$, then $\ell^{p_1,q_1} \subset \ell^{p_2,q_2}$.*

Finally, let us point out that the duality of ℓ^p spaces follows a familiar pattern for $1 \leq p < \infty$, that is, $(\ell^p)' = \ell^{p'}$, where $1/p + 1/p' = 1$. However, if $0 < p < 1$, then it can be shown that $(\ell^p)' = \ell^\infty$. Similarly, and as expected, we also have the following.

Proposition 1.18 *Let $1 \leq p, q < \infty$. Then $(L^p\ell^q)' = L^{p'}\ell^{q'}$. More precisely, for all $\mathbf{f} = (f_k)_{k\in\mathbb{Z}^d}$, any $F \in (L^p\ell^q)'$ can be expressed as*

$$F(\mathbf{f}) = \sum_{k\in\mathbb{Z}^d} \int_{\mathbb{R}^d} g_k(x) f_k(x)\, dx,$$

where $\mathbf{g} = (g_k)_{k\in\mathbb{Z}^d} \in L^{p'}\ell^{q'}$ and $\|F\|_{(L^p\ell^q)'} = \|\mathbf{g}\|_{L^{p'}\ell^{q'}}$.

1.2 Functional Analysis

Given two vector spaces $\mathscr{X}$ and $\mathscr{Y}$ over $\mathbb{C}$, a linear mapping $T : \mathscr{X} \to \mathscr{Y}$ is defined by the identity

$$T(\alpha x + \beta y) = \alpha T(x) + \beta T(y)$$

holding for all $x, y \in \mathscr{X}$ and $\alpha, \beta \in \mathbb{C}$. In the following, $(\mathscr{X}, \|\cdot\|_{\mathscr{X}})$ and $(\mathscr{Y}, \|\cdot\|_{\mathscr{Y}})$ will denote two abstract normed spaces.

1.2.1   Bounded Linear Operators: Fundamental Principles

Definition 1.19 The linear mapping $T : \mathscr{X} \to \mathscr{Y}$ is called *bounded* if there exists some constant $C \geq 0$ such that $\|T(x)\|_{\mathscr{Y}} \leq C\|x\|_{\mathscr{X}}$ for all $x \in \mathscr{X}$. The collection of all bounded linear operators $T : \mathscr{X} \to \mathscr{Y}$ is denoted by $\mathscr{B}(\mathscr{X}, \mathscr{Y})$.

It is a simple exercise to show that the following statements are all equivalent to $T \in \mathscr{B}(\mathscr{X}, \mathscr{Y})$:

(1) T is continuous on $\mathscr{X}$;
(2) T is continuous at 0;
(3) $\sup\{\|T(x)\|_{\mathscr{Y}} : \|x\|_{\mathscr{X}} \leq 1\} < \infty$;
(4) T is bounded on a dense subset of $\mathscr{X}$.

Thus, given a bounded linear operator $T : \mathscr{X} \to \mathscr{Y}$, we can talk about the *operator norm* of T:

$$\|T\| = \sup\{\|T(x)\|_{\mathscr{Y}} : \|x\|_{\mathscr{X}} \leq 1\} = \inf\{C \geq 0 : \|T(x)\|_{\mathscr{Y}} \leq C\|x\|_{\mathscr{X}}, \, \forall x \in \mathscr{X}\}.$$

In this way, we turn $\mathscr{B}(\mathscr{X}, \mathscr{Y})$ into a normed space. If the two spaces $\mathscr{X}$ and $\mathscr{Y}$ are assumed to be Banach, we have an important characterization for an operator T to belong to $\mathscr{B}(\mathscr{X}, \mathscr{Y})$.

Theorem 1.20 (Closed Graph Theorem) *The following are equivalent:*

(a) $T \in \mathscr{B}(\mathscr{X}, \mathscr{Y})$;
(b) *The graph* $\Gamma(T) = \{(x, T(x)) : x \in \mathscr{X}\}$ *of T is closed in* $\mathscr{X} \times \mathscr{Y}$;
(c) *If* $\|x_n\|_{\mathscr{X}} \to 0$ *and* $\|T(x_n) - y\|_{\mathscr{Y}} \to 0$ *as* $n \to \infty$, *then* $y = 0$.

Clearly, given a family of operators $(T_j)_{j \in \mathbb{J}} \subset \mathscr{B}(\mathscr{X}, \mathscr{Y})$, the uniform boundedness in the operator norm of $(T_j)_{j \in \mathbb{J}}$ implies the pointwise boundedness of the same family. The *Uniform Boundedness Principle* (or the Banach-Steinhaus theorem) asserts that the converse also holds if $\mathscr{X}$ is complete.

Theorem 1.21 (Banach-Steinhaus) *If $\mathscr{X}$ is Banach and for every $x \in \mathscr{X}$ the set* $\{T_j(x) : j \in \mathbb{J}\}$ *is bounded in $\mathscr{Y}$, then the set* $\{T_j : j \in \mathbb{J}\}$ *is bounded in $\mathscr{B}(\mathscr{X}, \mathscr{Y})$.*

Definition 1.22 The set $\mathscr{B}(\mathscr{X}, \mathbb{C})$ of bounded linear functionals $T : \mathscr{X} \to \mathbb{C}$ is denoted by $\mathscr{X}'$ and called the *dual space of* $\mathscr{X}$. Endowed with the operator norm, $\mathscr{X}'$ is a Banach space, regardless of whether $\mathscr{X}$ is complete or not.

Example 1.23 Let $\mathscr{X} = L^p(\mathbb{R}^d)$, $1 \leq p < \infty$. For a fixed $g \in L^{p'}(\mathbb{R}^d)$, we can define $T_g(f) = \int_{\mathbb{R}^d} f(x)g(x)\,dx$. Clearly, T_g is linear. T_g is also bounded; Hölder's inequality (1.3) immediately implies that $T_g \in (L^p)'$ with $\|T_g\|_{(L^p)'} \leq \|g\|_{L^{p'}}$. In fact,

$$\|f\|_{L^p} = \sup\{|T_g(f)| : \|g\|_{L^{p'}} \leq 1\},$$

effectively leading to the identification of the dual space $(L^p)'$ with $L^{p'}$. In particular, L^2 is self-dual, that is, $(L^2)' = L^2$.

1.2.2 Spaces of Smooth Functions

For $x = (x_1, x_2, \ldots, x_d) \in \mathbb{R}^d$ and a multi-index $\alpha = (\alpha_1, \alpha_2, \ldots, \alpha_d) \in (\mathbb{N} \cup \{0\})^d$ we write $|\alpha| = \sum_{i=1}^d \alpha_i$, $\alpha! = \prod_{i=1}^d \alpha_i!$, $x^\alpha = \prod_{i=1}^d x_i^{\alpha_i}$, and ∂^α for the higher order partial derivative operator $\prod_{i=1}^d \partial_{x_i}^{\alpha_i}$, where $\partial_{x_i}^{\alpha_i} = \partial^{\alpha_i}/\partial x_i^{\alpha_i}$. Moreover, if $\alpha = (\alpha_i)_{i=1}^d$, $\beta = (\beta_i)_{i=1}^d$ are two multi-indices, we can define the multi-indices $\alpha \pm \beta = (\alpha_i \pm \beta_i)_{i=1}^d$ and a partial order through $\alpha \leq \beta$ if and only if $\alpha_i \leq \beta_i$ for all $i \in \{1, 2, \ldots, d\}$. The inner product of two vectors $x, y \in \mathbb{R}^d$ is given by $x \cdot y = \sum_{i=1}^d x_i y_i$. Recall that the Euclidean norm of $x \in \mathbb{R}^d$ is then $|x| = (x \cdot x)^{1/2}$. Given two positive quantities a, b, we will write $a \lesssim b$ when $a \leq Cb$ for some constant C independent of a and b. If $a \lesssim b$ and $b \lesssim a$, we simply write $a \sim b$.

1.2.2.1 The Spaces $\mathscr{C}^k$, $\mathscr{C}^\infty$ and $\mathscr{C}_c^\infty$

Let $\mathbb{S}$ be an open subset of $\mathbb{R}^d$ and $k \in \mathbb{N} \cup \{0\}$.

Definition 1.24 The collection of all functions $f : \mathbb{S} \to \mathbb{C}$ such that $\partial^\alpha f$ is continuous for all multi-indices α such that $|\alpha| \leq k$ is denoted by $\mathscr{C}^k(\mathbb{S})$; while $\mathscr{C}^\infty(\mathbb{S}) = \bigcap_{k \geq 0} \mathscr{C}^k(\mathbb{S})$. $\mathscr{C}_c^\infty(\mathbb{S})$ (or $\mathscr{D}(\mathbb{S})$) denotes the collection of all functions $f \in \mathscr{C}^\infty(\mathbb{S})$ such that the support of f, supp(f), is compact and contained in $\mathbb{S}$. Here, we wrote

$$\text{supp}(f) = \overline{\{x \in \mathbb{R}^d : f(x) \neq 0\}}.$$

Note that given any two functions $f, g \in \mathscr{C}^{|\alpha|}(\mathbb{R}^d)$, the *product rule* holds:

$$\partial^\alpha(fg) = \sum_{\beta \leq \alpha} \frac{\alpha!}{\beta!(\alpha - \beta)!} (\partial^\beta f)(\partial^{\alpha - \beta} g).$$

The spaces $\mathscr{C}^k(\mathbb{S})$ and $\mathscr{C}^\infty(\mathbb{S})$ are examples of Frechét spaces (complete metric spaces) that are not normable. For example, the topology on $\mathscr{C}^\infty(\mathbb{S})$ is that of uniform convergence of a function and all its derivatives on compact sets; more precisely, the topology of $\mathscr{C}^\infty(\mathbb{S})$ is given by the seminorms

$$p_{j,m}(f) = \sup\{|\partial^\alpha f(x)| : x \in \mathbb{K}_j, |\alpha| \leq m\},$$

where $\mathbb{S} = \bigcup_{j=1}^\infty \mathbb{K}_j$, $\mathbb{K}_j$ compact and $\mathbb{K}_j \subset \mathbb{K}_{j+1}^\circ$. The notation $\mathbb{K}^\circ$ stands for the *interior* of the set $\mathbb{K}$.

Note now that $\mathscr{C}_c^\infty(\mathbb{S})$ is a nonempty subspace of $\mathscr{C}^\infty(\mathbb{S})$; the function ψ defined by $\psi(x) = e^{1/(|x|^2-1)}$ for $|x| < 1$ and $\psi(x) = 0$ for $|x| > 1$ is a classic example that belongs to $\mathscr{C}_c^\infty(\mathbb{R}^d)$. Inspired by the seminorms that determine the topology of $\mathscr{C}^\infty(\mathbb{S})$ it is tempting to impose on $\mathscr{C}_c^\infty(\mathbb{S})$ the topology given by the family of norms

$$\|f\|_m = \sup\{|\partial^\alpha f(x)| : x \in \mathbb{S}, |\alpha| \le m\}.$$

However, with this topology $\mathscr{C}_c^\infty(\mathbb{S})$ is incomplete. Thus, one prefers to give this space a topology that makes it complete, although not metrizable; more precisely, the convergence $f_k \to 0$ as $k \to \infty$ in $\mathscr{C}_c^\infty(\mathbb{S})$ means that there exists a compact subset $\mathbb{K} \subset \mathbb{S}$ such that for all k the supports of f_k are contained in $\mathbb{K}$ and for each multi-index α, $\partial^\alpha f_k$ converges uniformly to 0.

The properties of convolution immediately imply that if $f \in L^p(\mathbb{R}^d)$, $1 \le p \le \infty$, and $\varphi \in \mathscr{C}_c^\infty(\mathbb{R}^d)$, then $f * \varphi \in \mathscr{C}_c^\infty(\mathbb{R}^d)$, and from here is straightforward to show that $\mathscr{C}_c^\infty(\mathbb{R}^d)$ is dense in $L^p(\mathbb{R}^d)$.

1.2.2.2 The Space $\mathscr{S}$

In analogy to the previous definitions, we can define the Schwartz space $\mathscr{S}(\mathbb{R}^d)$ of rapidly decreasing smooth functions as follows.

Definition 1.25 We say that $f \in \mathscr{S}(\mathbb{R}^d)$ (or simply $f \in \mathscr{S}$) if $f \in \mathscr{C}^\infty(\mathbb{R}^d)$ and

$$\|f\|_{N,\alpha} = \sup\{(1 + |x|)^N |\partial^\alpha f(x)| : x \in \mathbb{R}^d\} < \infty$$

for all nonnegative integers N and multi-indices α.

Equivalently, the complete topology of $\mathscr{S}$ can be given through the family of seminorms

$$p_{\alpha,\beta}(f) = \sup\{|x^\alpha \partial^\beta f(x)| : x \in \mathbb{R}^d\}.$$

Using the product rule and the properties of convolution, it is not hard to prove that for all f, g in $\mathscr{S}$, we have fg and $f * g$ also in $\mathscr{S}$. It is also easy to see that we have the strict inclusions

$$\mathscr{C}_c^\infty(\mathbb{R}^d) \subset \mathscr{S}(\mathbb{R}^d) \subset \mathscr{C}^\infty(\mathbb{R}^d).$$

Moreover, $\mathscr{C}_c^\infty$ is dense in $\mathscr{S}$ and $\mathscr{S} \subset L^p$ for all $1 \le p \le \infty$. For the last inclusion, simply note that if $p < \infty$ and $f \in \mathscr{S}(\mathbb{R}^d)$, then

$$\|f\|_{L^p}^p = \int_{\mathbb{R}^d} |f(x)|^p \, dx \lesssim \int_{\mathbb{R}^d} (1 + |x|)^{-d-1} \, dx < \infty.$$

1.2.3 Distributions

Definition 1.26 The dual space of $\mathscr{D}$, that is, $\mathscr{D}' = \mathscr{B}(\mathscr{D}, \mathbb{C})$, is called *the space of distributions*. The dual space of $\mathscr{S}$, that is, $\mathscr{S}' = \mathscr{B}(\mathscr{S}, \mathbb{C})$, is called *the space of tempered distributions*.

Note that a functional $u \in \mathscr{S}'$ if and only if there exist $N, M \in \mathbb{N}$ such that

$$|u(f)| \lesssim \sum_{|\alpha| \leq N} \sum_{|\beta| \leq M} p_{\alpha,\beta}(f) \text{ for all } f \in \mathscr{S}.$$

A functional u belongs to $\mathscr{D}'$ if for each compact set $\mathbb{K} \subset \mathbb{R}^d$ there exist a nonnegative integer m and a positive constant C (both dependent on $\mathbb{K}$) such that for all $f \in \mathscr{C}_c^\infty(\mathbb{R}^d)$ with f supported in $\mathbb{K}$, we have $|u(f)| \leq C \|f\|_m$.

From now on, if $u \in \mathscr{S}'$ (or $\mathscr{D}'$) and $f \in \mathscr{S}$ (or $\mathscr{D}$), we will write $\langle u, f \rangle$ for $u(f)$. We can endow $\mathscr{S}'$ (or $\mathscr{D}'$) with the weak convergence topology: we say, for example, that a sequence of distributions $u_j \in \mathscr{S}'$ converges to $u \in \mathscr{S}'$ if $\langle u_j, f \rangle$ converges to $\langle u, f \rangle$ for all $f \in \mathscr{S}$

Since $\mathscr{D} \subset \mathscr{S}$, we have $\mathscr{S}' \subset \mathscr{D}'$. This inclusion is, of course, strict.

Example 1.27 Let $\varphi \in L^1_{\text{loc}}(\mathbb{R}^d)$, and construct a linear functional u_φ on $\mathscr{D}$, by letting

$$\langle u_\varphi, f \rangle := \int_{\mathbb{R}^d} \overline{\varphi(x)} f(x) \, dx.$$

Clearly, if $\text{supp}(f) \subset \mathbb{K}$ (compact), we immediately have $|\langle u_\varphi, f \rangle| \leq C(\varphi, \mathbb{K}) \|f\|_0$, where $C(\varphi, \mathbb{K}) = \|\varphi \chi_{\mathbb{K}}\|_{L^1}$. Thus $u_\varphi \in \mathscr{D}'$. However, for the specific choice $\varphi(x) = e^{|x|^2}$ (which is obviously in $L^1_{\text{loc}}(\mathbb{R}^d)$), it is easy to see that u_φ does not belong to $\mathscr{S}'$.

Nevertheless, a canonical way in which one can create a tempered distribution u_φ is to require that the locally integrable function φ is also tempered at infinity.

Example 1.28 Let $\varphi \in L^1_{\text{loc}}(\mathbb{R}^d)$ be such that $|\varphi(x)| \lesssim (1+|x|)^N$ for $|x|$ sufficiently large and some nonnegative integer N. It is straightforward to show in this case that $|\langle u_\varphi, f \rangle| \lesssim p_{N+d+1,0}(f)$ for all $f \in \mathscr{S}$. Thus $u_\varphi \in \mathscr{S}'$.

Another canonical way of obtaining a tempered distribution $u_\varphi \in \mathscr{S}'$ is to require that the locally integrable function φ is p-integrable.

Example 1.29 Let $\varphi \in L^p$, $1 \leq p \leq \infty$. Hölder's inequality (1.1) and the inclusion $\mathscr{S} \subset L^p$ immediately give $|\langle u_\varphi, f \rangle| \leq \|\varphi\|_{L^p} \|f\|_{L^{p'}} \lesssim \|\varphi\|_{L^p} \|f\|_{N,\alpha}$, for appropriate indices N, α. Thus $u_\varphi \in \mathscr{S}'$.

Finally, let us consider the Dirac distribution δ_0 concentrated at the origin.

Example 1.30 For $f \in \mathscr{S}$, define $\langle \delta_0, f \rangle = f(0)$. δ_0 is (trivially) a tempered distribution, since $|\langle \delta_0, f \rangle| \le \|f\|_{0,0}$.

Remark 1.31 We close this subsection with a remark about the definition of the distribution u_φ in Example 1.27. Clearly,

$$\langle u_\varphi, f \rangle = \langle f, \varphi \rangle_{L^2} = \overline{\langle \varphi, f \rangle_{L^2}}.$$

In particular,

$$|\langle u_\varphi, f \rangle| = |\langle \varphi, f \rangle_{L^2}|.$$

1.2.4 Translation, Dilation, and Modulation of a Distribution

The usual operations on a smooth function have a counterpart in the realm of distributions. We can talk, for example, about the translation, dilation, or derivative of a distribution.

To set ideas, let $u \in \mathscr{D}'$ and $f \in \mathscr{D}$; let also $x, y, \xi \in \mathbb{R}^d$, $\epsilon > 0$, and $\varphi \in \mathscr{C}^\infty(\mathbb{R}^d)$. Then, the *translation, dilation, and modulation operators* (acting on functions) are defined as follows:

(a) (Translation, or time-shift, by y) $T_y f(x) = f(x - y)$;
(b) (L^1-normalized ϵ-dilation) $D_\epsilon f(x) = \epsilon^{-d} f(\epsilon^{-1} x)$;
(c) (Modulation, or frequency-shift, by ξ) $M_\xi f(x) = e^{2\pi i \xi \cdot x} f(x)$.

We re-emphasize that the definition that we pick here (in (b)) for the dilation operator is the one normalized in L^1; the normalization factor ϵ^{-d} can be of course omitted, but in this case one needs to adjust accordingly in the computation of the L^1 norm of the dilation operator. The *time-frequency shifts* $T_y M_\xi$ and $M_\xi T_y$ will play an important role for us since they are at the core of the definition of modulation spaces. A simple calculation shows that T_y and M_ξ *almost commute*:

$$T_y M_\xi = e^{-2\pi i y \cdot \xi} M_\xi T_y. \tag{1.9}$$

In particular, the space and frequency shifts (by y and ξ, respectively) do commute if and only if $y \cdot \xi \in \mathbb{Z}$. Moreover, time-frequency shifts interact nicely with the Fourier and short-time Fourier transforms; see Sect. 1.3.

Definition 1.32 We can define the operations of translation, dilation, and modulation, respectively, of a distribution u in a natural way, by "moving" the operations on the test function f on which u is acting:

(A) $\langle T_y u, f \rangle = \langle u, T_{-y} f \rangle$;
(B) $\langle D_\epsilon u, f \rangle = \langle u, f(\epsilon \cdot) \rangle$;
(C) $\langle M_\xi u, f \rangle = \langle u, M_{-\xi} f \rangle$.

In fact, the modulation operation can be viewed as a particular case of the operation of *multiplication by a smooth function* φ *of a distribution* u:

$$\langle \varphi u, f \rangle := \langle u, \overline{\varphi} f \rangle.$$

The connection between modulation and multiplication by a smooth function is the following: given a test function f, we have $M_\xi f = \varphi_\xi f$, where $\varphi_\xi(x) = e^{2\pi i \xi \cdot x}$; then, for the distribution u, we would similarly define $M_\xi u = \varphi_\xi u$. The definitions above are natural, in the sense that if the distribution that we consider is coming from a smooth function ψ, that is of the form u_ψ, we would obtain $M_\xi u_\psi = u_{\varphi_\xi \psi}$.

Definition 1.33 We define the reflection and conjugation of a distribution u, respectively, by

$$\langle \tilde{u}, f \rangle = \langle u, \tilde{f} \rangle, \quad \langle \bar{u}, f \rangle = \langle u, \bar{f} \rangle,$$

where $\tilde{f}(x) = f(-x)$, $\bar{f}(x) = \overline{f(x)}$, and f is any test function.

Example 1.34 We let $\delta_y := T_y \delta_0$, where δ_0 is the Dirac distribution concentrated at the origin; we call δ_y the Dirac distribution concentrated at y. By definition, given some test function f, we have

$$\langle \delta_y, f \rangle = \langle \delta_0, T_{-y} f \rangle = T_{-y} f(0) = f(y).$$

Example 1.35 We specify the action of the distribution $M_\xi \delta_y$. Given some test function f, we can write

$$\langle M_\xi \delta_y, f \rangle = \langle \delta_y, M_{-\xi} f \rangle = (M_{-\xi} f)(y) = e^{-2\pi i \xi \cdot y} f(y).$$

In particular, $\langle M_\xi \delta_0, f \rangle = f(0) = \langle \delta_0, f \rangle$; thus $M_\xi \delta_0 = \delta_0$ for all $\xi \in \mathbb{R}^d$.

1.2.5 Derivative of a Distribution

Definition 1.36 We define the partial derivative of a distribution u via

$$\langle \partial^\alpha u, f \rangle = (-1)^{|\alpha|} \langle u, \partial^\alpha f \rangle,$$

where f is any test function.

Let us consider a few examples that illustrate this concept.

Example 1.37 Let $\varphi \in \mathscr{C}^1(\mathbb{R}^d)$. Since $\varphi \in L^1_{\text{loc}}(\mathbb{R}^d)$, we can define u_φ as an element of $\mathscr{D}'$; recall that

$$\langle u_\varphi, f \rangle = \int_{\mathbb{R}^d} \overline{\varphi(x)} f(x) \, dx.$$

It is easy to see that for all multi-indices α with $|\alpha| = 1$ we have $\partial^\alpha u_\varphi = u_{\partial^\alpha \varphi}$. The right-hand side of this equality is meaningful since $\partial^\alpha \varphi \in L^1_{\text{loc}}(\mathbb{R}^d)$. This example also points out that it is justified to abuse the notation and simply write φ to denote also the distribution u_φ; the context should make it clear whether we talk about a function or a distribution.

Example 1.38 The "principal value distribution" p.v.$\frac{1}{x}$ is defined by

$$\left\langle \text{p.v.}\frac{1}{x}, f \right\rangle = \lim_{\epsilon \to 0} \int_{|x| > \epsilon} \frac{f(x)}{x}\, dx.$$

Splitting the integral over the intervals $(-\infty, -\epsilon)$ and (ϵ, ∞) and using integration by parts we get

$$\left\langle \text{p.v.}\frac{1}{x}, f \right\rangle = -\int_{\mathbb{R}} \ln|x| f'(x)\, dx + (f(-\epsilon) - f(\epsilon)) \ln \epsilon.$$

But $|(f(-\epsilon) - f(\epsilon)) \ln \epsilon| \leq 2\|f'\|_{L^\infty} \epsilon |\ln \epsilon| \to 0$ as $\epsilon \to 0$. Therefore, in distributional sense, $(\ln|x|)' = \text{p.v.}\frac{1}{x}$.

Example 1.39 Let $\varphi = \chi_{\mathbb{R}\setminus\{0\}} \in L^1_{\text{loc}}(\mathbb{R})$. We claim that, as distributions, $\varphi' = 0$; that is, for all $f \in \mathscr{D}$ we have $\langle u'_\varphi, f \rangle = 0$. Indeed, if we let $a > 0$ be such that the support of f is contained in $(-a, a)$, we see that

$$\langle u'_\varphi, f \rangle = -\langle u_\varphi, f' \rangle = -\int_{\mathbb{R}} f'(x)\, dx = f(-a) - f(a) = 0.$$

Example 1.40 Let now $\varphi = \chi_{(0,\infty)}$. A similar computation to the one above gives $\varphi' = \delta_0$, that is, for all $f \in \mathscr{D}$ we have $\langle u'_\varphi, f \rangle = f(0)$.

Example 1.41 Finally, let $\varphi(x) = \text{sign } x$, that is, $\varphi(x) = 1$ for $x > 0$ and $\varphi(x) = -1$ for $x < 0$. We have that $\varphi' = 2\delta_0$. Indeed, as in Example 1.39, we compute

$$\langle u'_\varphi, f \rangle = -\langle u_\varphi, f' \rangle = \int_{-a}^{0} f'(x)\, dx - \int_{0}^{a} f'(x)\, dx = 2f(0).$$

Example 1.42 By definition, the derivative of the Dirac distribution concentrated at y is given by $\langle \delta'_y, f \rangle = -\langle \delta_y, f' \rangle = -f'(y)$. Incidentally, δ'_y is an example of a *distribution that is not a function*; see Example 1.37.

1.2.6   Convolution Between a Distribution and a Test Function

Definition 1.43 The convolution between a distribution $u \in \mathscr{D}'$ and a test function $f \in \mathscr{D}$ is *a function*, defined as follows:

$$u * f(x) = \langle u, \widetilde{T_x f} \rangle = \langle u, f(x - \cdot) \rangle.$$

The following properties of convolution follow straightforwardly from the definition:

(1) $T_y(u * f) = T_y u * f = u * T_y f$;
(2) $\partial^\alpha(u * f) = \partial^\alpha u * f = u * \partial^\alpha f$ for all multi-indices α;
(3) $u * f \in \mathscr{C}^\infty$.

Moreover, *if $u \in \mathscr{S}'$ and $f \in \mathscr{S}$, then $u * f$ is a tempered function*, that is, $|u * f(x)| \lesssim (1 + |x|)^N$ for some positive integer N. Let us briefly indicate why. By the definition of $\mathscr{S}'$, we know that there exists N, k nonnegative integers such that for all $g \in \mathscr{S}$ we have

$$|\langle u, g \rangle| \lesssim \sup_{|\alpha| \leq k} \|g\|_{N,\alpha}.$$

Therefore, using the definition of convolution and of $\mathscr{S}$, we get

$$|u * f(x)| = |\langle u, f(x - \cdot) \rangle| \lesssim \sup\{(1 + |y|)^N |\partial_y^\alpha f(x - y)| : |\alpha| \leq k, y \in \mathbb{R}^d\}$$

$$\lesssim \sup\{(1 + |x - y|)^N (1 + |x|)^N |\partial_y^\alpha f(x - y)| : |\alpha| \leq k, y \in \mathbb{R}^d\}$$

$$\lesssim (1 + |x|)^N.$$

Finally, let us remark that since the convolution $u * f$ of $u \in \mathscr{S}'$ and $f \in \mathscr{S}$ is a tempered function at infinity, we can extend it to a tempered distribution in $\mathscr{S}'$; writing $u * f$ also for this distribution, it is easy to see that

$$\langle u * f, g \rangle = \langle u, \tilde{f} * g \rangle.$$

1.3 Fourier Analysis

The two basic concepts that will be fundamental for our purposes are the Fourier transform and its short-time version. We introduce them here and list some of their simple, but essential, properties.

1.3.1 The Fourier Transform

1.3.1.1 Definition and Basic Properties

Definition 1.44 Let $\xi \in \mathbb{R}^d$ and $f \in L^1(\mathbb{R}^d)$. The *Fourier transform of f* is denoted by $\mathscr{F}(f)$, or simply $\hat{f}$, and it is defined by

$$\widehat{f}(\xi) = \int_{\mathbb{R}^d} f(x) e^{-2\pi i x \cdot \xi} \, dx. \tag{1.10}$$

Note that the integral in (1.10) is well defined and $\widehat{f}$ becomes a continuous function of ξ. Moreover, $\mathscr{F}$ is a bounded linear operator from L^1 to L^∞ and the operator norm $\|\mathscr{F}\|_{\mathscr{B}(L^1, L^\infty)} \leq 1$. In fact, $\mathscr{F}$ maps L^1 into $\mathscr{C}_0$, *the subspace of continuous functions that vanish at infinity*. This is the content of the next result.

Lemma 1.45 (Riemann-Lebesgue) *If $f \in L^1(\mathbb{R}^d)$, then $\lim_{|\xi| \to \infty} \widehat{f}(\xi) = 0$.*

Proof Let $f \in \mathscr{C}_c^\infty(\mathbb{R}^d)$ and $\xi \neq 0$. Integrating by parts twice gives that $\widehat{f}(\xi) = -|\xi|^{-2} \widehat{\Delta f}(\xi)$, where we denoted by Δ the Laplacian $\Delta f = \sum_{i=1}^{d} \partial_{x_i}^2 f$. Since $\Delta f \in L^1(\mathbb{R}^d)$, we know that $\widehat{\Delta f} \in L^\infty(\mathbb{R}^d)$, thus $|\widehat{f}(\xi)| \lesssim |\xi|^{-2} \to 0$ as $|\xi| \to \infty$. Now, given a general $f \in L^1(\mathbb{R}^d)$, we simply approximate it with a sequence $(f_k) \subset \mathscr{C}_c^\infty(\mathbb{R}^d)$ in the L^1 norm. This gives $\|\widehat{f_k} - \widehat{f}\|_{L^\infty} \to 0$ as $k \to \infty$. Since we already proved that $\widehat{f_k}(\xi) \to 0$ as $|\xi| \to \infty$ for all k, we immediately get $\widehat{f}(\xi) \to 0$ as $|\xi| \to \infty$. □

The Fourier transform interacts nicely with the translation, dilation, modulation, reflection, conjugation, differentiation, and convolution operations.

Proposition 1.46 *Let $f, g \in L^1(\mathbb{R}^d)$, $y, \xi, \zeta \in \mathbb{R}^d$, $\epsilon > 0$, and α a multi-index. The following identities hold.*

(a) $\widehat{T_y f} = M_{-y} \widehat{f}$ and $\widehat{M_\xi f} = T_\xi \widehat{f}$;
(b) $\widehat{T_y M_\xi f} = M_{-y} T_\xi \widehat{f} = e^{-2\pi i y \cdot \xi} T_\xi M_{-y} \widehat{f}$;
(c) $\widehat{D_\epsilon f}(\zeta) = \widehat{f}(\epsilon \zeta)$;
(d) $\mathscr{F}(\tilde{f}) = \widetilde{\mathscr{F}(f)}$;
(e) $\mathscr{F}(\bar{f}) = \overline{\widetilde{\mathscr{F}(f)}}$;
(f) $\widehat{\partial^\alpha f}(\zeta) = (2\pi i \zeta)^\alpha \widehat{f}$ and $\partial^\alpha \widehat{f} = (-2\pi i)^{|\alpha|} \widehat{x^\alpha f(x)}$;
(g) $\widehat{f * g} = \widehat{f} \widehat{g}$.
(h) *Moreover, if U is a $d \times d$ orthogonal matrix, then $\widehat{f \circ U}(\zeta) = \widehat{f}(U\zeta)$.*

Let us point out how (a) and (g) are obtained. We assume that in our calculations all the integrals are over $\mathbb{R}^d$. The identities in (a) follow from the following simple manipulations:

$$\widehat{T_y f}(\zeta) = \int T_y f(x) e^{-2\pi i x \cdot \zeta} \, dx = \int f(x) e^{-2\pi i (x+y) \cdot \zeta} \, dx = e^{-2\pi i y \cdot \zeta} \widehat{f}(\zeta)$$
$$= M_{-y} \widehat{f}(\zeta),$$

$$\widehat{M_\xi f}(\zeta) = \int M_\xi f(x) e^{-2\pi i x \cdot \zeta} \, dx = \int f(x) e^{-2\pi i (\zeta - \xi) \cdot \zeta} \, dx = \widehat{f}(\zeta - \xi) = T_\xi \widehat{f}(\zeta).$$

(g) is a nice application of Fubini's Theorem 1.7. We have the following sequence of equalities:

$$\widehat{f * g}(\xi) = \int (f * g)(x)e^{-2\pi i x \cdot \xi}\, dx = \int \left(\int f(y)g(x-y)\, dy\right)e^{-2\pi i x \cdot \xi}\, dx$$

$$= \int f(y)e^{-2\pi i y \cdot \xi}\left(\int g(x-y)e^{-2\pi i (x-y) \cdot \xi}\, dx\right)dy$$

$$= \int f(y)e^{-2\pi i y \cdot \xi}\, dy \int g(x)e^{-2\pi i x \cdot \xi}\, dx = \widehat{f}(\xi)\widehat{g}(\xi).$$

The other identities follow from equally simple considerations and are left as exercises. For example, (b) is a direct consequence of (a) and the almost commutation property of time-frequency shifts (1.9), see Sect. 1.2.4.

An immediate corollary of Proposition 1.46 is that the Fourier transform $\mathscr{F}$ *is a continuous mapping from $\mathscr{S}$ to $\mathscr{S}$*. This claim follows by estimating any semi-norm $p_{\alpha,\beta}(\widehat{f})$ by another semi-norm $p_{\alpha',\beta'}(f)$ of a function $f \in \mathscr{S}$. Note first that, given $f \in \mathscr{S}$, we have $\widehat{f} \in \mathscr{C}^\infty$ and, by Proposition 1.46,

$$p_{\alpha,\beta}(\widehat{f}) = \sup\{|\xi^\alpha \partial^\beta \widehat{f}(\xi)| : \xi \in \mathbb{R}^d\}$$

$$= (2\pi i)^{-|\alpha|} \sup\{|\partial^\alpha \widehat{((-2\pi i x)^\beta f(x))}(\xi)| : \xi \in \mathbb{R}^d\}.$$

Thus, for a fixed $\xi \in \mathbb{R}^d$, and $N > d$, we have

$$|\xi^\alpha \partial^\beta \widehat{f}(\xi)| \lesssim \sup\{|\partial^\alpha(x^\beta \widehat{f}(x))|(1+|x|)^N : x \in \mathbb{R}^d\}$$

$$\lesssim \sup\{(1+|x|)^{N+|\alpha|}|\partial^\gamma f(x)| : |\gamma| \leq |\beta|, x \in \mathbb{R}^d\} \lesssim \|f\|_{N+|\alpha|,\beta}.$$

In fact, the mapping $\mathscr{F} : \mathscr{S} \to \mathscr{S}$ *is an isomorphism*, that is, $\mathscr{F}$ is a bijection and its inverse $\mathscr{F}^{-1}$ is also continuous from $\mathscr{S}$ to $\mathscr{S}$. The argument that proves this uses another nice fact: the standard Gaussian function is left unchanged by the Fourier transform. We show this below, along with a few other simple calculations of the Fourier transform.

Example 1.47 Let $x \in \mathbb{R}^d$ and $\varphi_d(x) = e^{-\pi|x|^2}$. We claim that $\widehat{\varphi}_d = \varphi_d$. Note first that it is enough to obtain this equality for $d = 1$. Indeed, assuming the one-dimensional case holds true, the general case follows by separating the one-dimensional variables; that is, if $x = (x_1, x_2, \ldots, x_d), \xi = (\xi_1, \xi_2, \ldots, \xi_d)$, then

$$\widehat{\varphi}_d(\xi) = \int_{\mathbb{R}^d} e^{-\pi|x|^2}e^{-2\pi i x \cdot \xi}\, dx = \int_{\mathbb{R}^d} e^{-\pi \sum_{j=1}^d x_j^2}e^{-2\pi i \sum_{j=1}^d x_j \xi_j}\, dx_1 \cdots dx_d$$

$$= \prod_{j=1}^d \int_{\mathbb{R}} e^{-\pi x_j^2}e^{-2\pi i x_j \xi_j}\, dx_j = \prod_{j=1}^d \widehat{\varphi}_1(\xi_j) = \prod_{j=1}^d e^{-\pi \xi_j^2} = e^{-\pi|\xi|^2} = \varphi_d(\xi).$$

The computation of $\widehat{\varphi_1}$ follows from an appropriate completion of squares. Let $x, \xi \in \mathbb{R}$; we have

$$\widehat{\varphi_1}(\xi) = e^{-\pi\xi^2} \int_{\mathbb{R}} e^{-\pi(x+i\xi)^2} \, dx = \varphi_1(\xi) \int_{\mathbb{R}} e^{-\pi x^2} \, dx = \varphi_1(\xi).$$

Alternately, we could observe that φ_1 satisfies the first-order ordinary differential equation $\varphi_1'(x) + 2\pi x \varphi_1(x) = 0$. Therefore, by applying the Fourier transform to this equality and using Proposition 1.46 we see that $2\pi\xi\widehat{\varphi_1}(\xi) + \widehat{\varphi_1}'(\xi) = 0$ for all $\xi \in \mathbb{R}$. Thus,

$$\widehat{\varphi_1}(\xi) = \widehat{\varphi_1}(0)e^{-\pi\xi^2} = \varphi_1(\xi).$$

Example 1.48 Let $x \in \mathbb{R}$ and $f(x) = e^{-|x|}$. We claim that $\widehat{f}(\xi) = \frac{1}{1+4\pi^2\xi^2}$. Indeed, using Euler's formula and the fact that $x \mapsto e^{-|x|}\cos(2\pi x\xi)$ and $x \mapsto e^{-|x|}\sin(2\pi x\xi)$ are even, respectively odd, functions we obtain

$$\widehat{f}(\xi) = 2 \int_0^\infty e^{-x} \cos(2\pi x\xi) \, dx$$

$$= \frac{e^{-x}\left(2\pi\xi\sin(2\pi x\xi) - \cos(2\pi x\xi)\right)}{1 + 4\pi^2\xi^2}\Big|_{x=0}^{x=\infty} = \frac{1}{1 + 4\pi^2\xi^2}.$$

A slightly more involved computation proves that if $x \in \mathbb{R}^d$ and f is defined as above, then

$$\widehat{f}(\xi) = c_d(1 + 4\pi^2|\xi|^2)^{-(d+1)/2},$$

for some positive constant c_d.

Example 1.49 Let now $f_1 = \chi_{[0,1]}$. Then $\widehat{f_1}(\xi) = \frac{1-e^{-2\pi i\xi}}{2\pi i\xi}$. As in Example 1.47, if we separate variables out, letting $f_d = \chi_{[0,1]^d}$ we get

$$\widehat{f_d}(\xi_1, \xi_2, \ldots, \xi_d) = \prod_{j=1}^d \frac{1 - e^{-2\pi i\xi_j}}{2\pi i\xi_j}.$$

Similarly, if $g_d = \chi_{[-1,1]^d}$, we compute

$$\widehat{g_d}(\xi_1, \xi_2, \ldots, \xi_d) = \prod_{j=1}^d \frac{\sin(2\pi\xi_j)}{\pi\xi_j}.$$

Definition 1.50 Let $f \in \mathscr{S}(\mathbb{R}^d)$, and define $f^{\vee} = \widetilde{\mathscr{F}(f)}$; that is, for $x \in \mathbb{R}^d$,

$$f^{\vee}(x) = \widehat{f}(-x) = \int_{\mathbb{R}^d} f(\xi) e^{2\pi i x \cdot \xi} \, d\xi.$$

We claim that the operation $\vee$ is exactly the *Fourier inversion* :

$$\mathscr{F}^{-1}(f) = f^{\vee}.$$

We prove first that $f = (\widehat{f})^{\vee}$, that is,

$$f(x) = \int_{\mathbb{R}^d} \widehat{f}(\xi) e^{2\pi i x \cdot \xi} \, d\xi.$$

Fix $x \in \mathbb{R}^d$. Start by noticing that if $f, \varphi \in \mathscr{S}(\mathbb{R}^d)$, Fubini's Theorem 1.7 immediately gives

$$\int_{\mathbb{R}^d} f(y) \widehat{\varphi}(y) \, dy = \int_{\mathbb{R}^d} \widehat{f}(\xi) \varphi(\xi) \, d\xi. \tag{1.11}$$

Let us apply (1.11) by replacing $\varphi(y)$ with $\epsilon^{-d} M_x D_{\epsilon^{-1}} \varphi_d(y) = e^{2\pi i y \cdot x} \varphi_d(\epsilon y) = e^{2\pi i y \cdot x} e^{-\pi |\epsilon y|^2}$, where $\epsilon > 0$ is arbitrary. Further using Proposition 1.46 (a) in (1.11) and Example 1.47, we see that

$$\int_{\mathbb{R}^d} f(x + \epsilon y) \varphi_d(y) \, dy = \int_{\mathbb{R}^d} f(x + \epsilon y) \widehat{\varphi_d}(y) \, dy = \int_{\mathbb{R}^d} \widehat{f}(\xi) \varphi_d(\epsilon \xi) e^{2\pi i x \cdot \xi} \, d\xi.$$

By applying now Lebesgue's Dominated Convergence Theorem 1.5 and letting $\epsilon \to 0$ gives

$$f(x) \widehat{\varphi_d}(0) = \varphi_d(0) \int_{\mathbb{R}^d} \widehat{f}(\xi) e^{2\pi i x \cdot \xi} \, d\xi,$$

which is exactly what we wanted to prove. Moreover, by replacing f with the reflection $\widetilde{f}$, we also get $\mathscr{F}(f^{\vee}) = f$, thus proving the invertibility of $\mathscr{F}$ on $\mathscr{S}$. The definition of $\vee$ and the already proven continuity of $\mathscr{F}$ show that $\mathscr{F}^{-1}$ is also continuous on $\mathscr{S}$.

A close inspection of the argument outlined above shows that we could repeat it under the weaker assumptions $f \in L^1(\mathbb{R}^d)$ and $\widehat{f} \in L^1(\mathbb{R}^d)$ and get that, almost everywhere, $f = (\widehat{f})^{\vee}$. In fact, the Fourier inversion formula can be extended to L^2. This is a consequence of two beautiful identities due to Parseval and Plancherel.

Proposition 1.51 *Let $f, g \in \mathscr{S}(\mathbb{R}^d)$. We have the following.*

(a) *(**Parseval's identity**) $\int_{\mathbb{R}^d} f(x) \overline{g(x)} \, dx = \int_{\mathbb{R}^d} \widehat{f}(\xi) \overline{\widehat{g}(\xi)} \, d\xi$;*
(b) *(**Plancherel's identity**) $\|f\|_{L^2} = \|\widehat{f}\|_{L^2} = \|f^{\vee}\|_{L^2}$.*

In the language of L^2-inner products, we have shown that Parseval's identity is nothing but an *orthogonality relation of the Fourier transform*:

$$\langle f, g \rangle_{L^2} = \langle \hat{f}, \hat{g} \rangle_{L^2}.$$

Note also that, using the density of $\mathscr{S}$ in L^2, Plancherel's identity proves that $\mathscr{F}$ can be extended as an isomorphism from L^2 to L^2.

Proof Plancherel's identity is a consequence of Parseval's identity; simply let $f = g$ in (a). Parseval's identity follows from the Fourier inversion formula and the identity (1.11) as follows:

$$\int_{\mathbb{R}^d} f(x)\overline{g(x)}\, dx = \int_{\mathbb{R}^d} f(x)\widehat{\bar{g}^\vee}(x)\, dx$$

$$= \int_{\mathbb{R}^d} \hat{f}(\xi)\bar{g}^\vee(\xi)\, d\xi = \int_{\mathbb{R}^d} \hat{f}(\xi)\hat{\bar{g}}(-\xi)\, d\xi$$

$$= \int_{\mathbb{R}^d} \hat{f}(\xi)\overline{\hat{g}(\xi)}\, d\xi.$$

$\square$

The action of $\mathscr{F}$ on $L^p, 1 \le p \le 2$, follows by interpolating Plancherel's inequality (identity) $\|\hat{f}\|_{L^2} \le \|f\|_{L^2}$ and the trivial inequality $\|\hat{f}\|_{L^1} \le \|f\|_{L^\infty}$. The outcome is *the Hausdorff-Young inequality*.

Proposition 1.52 (Hausdorff-Young) *Let* $f \in L^p$ *with* $1 \le p \le 2$. *Then,*

$$\|\hat{f}\|_{L^{p'}} \le \|f\|_{L^p}.$$

A sharp version of this inequality is due to Beckner [3]:

$$\|\hat{f}\|_{L^{p'}} \le \left(\frac{p^{1/p}}{p'^{1/p'}} \right)^{1/2} \|f\|_{L^p}.$$

The weighted extension of Proposition 1.52 is known as *Pitt's inequality* [196].

Proposition 1.53 (Pitt) *Let* $1 \le p \le 2$ *and* $0 \le \alpha < d/p'$. *Then,*

$$\||x|^{-\alpha} \widehat{f}\|_{L^{p'}(\mathbb{R}^d)} \le \||x|^\alpha f\|_{L^p(\mathbb{R}^d)}.$$

While we know that, in general, there is no embedding between L^p spaces on $\mathbb{R}^d$, an additional assumption on the support of the Fourier transform changes that. Let $\xi_0 \in \mathbb{R}^d$ and the compact set $\mathbb{K} \subset \mathbb{R}^d$ be fixed. Following the notation used in [166], we introduce the following space of functions.

Definition 1.54 Let $0 < p \leq \infty$. We say that $f \in PW^p(\mathbb{K}; \xi_0)$ if $f \in L^p(\mathbb{R}^d)$ and supp $(\widehat{f}) \subset \xi_0 + \mathbb{K}$. In particular, when $\xi_0 = 0$ we simply write $PW^p(\mathbb{K})$ for $PW^p(\mathbb{K}; 0)$.

Note that, by Theorem 1.71, $PW^p(\mathbb{K}; \xi_0)$ consists of real analytic functions. With this notation, the *Nikol'skij-Triebel inequality* holds; see [227, Proposition 1.3.2].

Proposition 1.55 (Nikol'skij-Triebel) *Let $\mathbb{K} \subset \mathbb{R}^d$ be a compact set and $0 < p \leq q \leq \infty$. If $f \in PW^p(\mathbb{K})$, then there exists a constant $C = C(\operatorname{diam} \mathbb{K}, p)$ such that* $\|f\|_{L^q} \leq C\|f\|_{L^p}$.

Here, $\operatorname{diam} \mathbb{K} := \sup\{|x - y| : x, y \in \mathbb{K}\}$ denotes the (finite) diameter of $\mathbb{K}$. An immediate consequence of Nikol'skij-Triebel's inequality is the following convolution inequality, see [227, Proposition 1.5.3].

Proposition 1.56 *Let $\mathbb{K}_1, \mathbb{K}_2 \subset \mathbb{R}^d$ be compact sets and $0 < p \leq 1$. If $f \in PW^p(\mathbb{K}_1)$ and $g \in PW^p(\mathbb{K}_2)$, then there exists a constant $C = C(\operatorname{diam} \mathbb{K}_1, \operatorname{diam} \mathbb{K}_2, p)$ such that $\||f| * |g|\|_{L^p} \leq C\|f\|_{L^p}\|g\|_{L^p}$.*

Proof Given $x \in \mathbb{R}^d$, let $h_x := f\widetilde{T_x g}$, that is, $h_x(y) = f(y)g(x - y)$. Then, for a.e. $x \in \mathbb{R}^d$, $h_x \in L^p$. Using now Proposition 1.46, we have $\widehat{h_x} = \widehat{f} * (\widetilde{M_{-x}\widehat{g}})$. Therefore, the hypothesis tells us that supp $(\widehat{h_x}) \subset \mathbb{K}_1 - \mathbb{K}_2$, where $\mathbb{K}_1 - \mathbb{K}_2$ denotes the compact set $\{x_1 - x_2 | x_j \in \mathbb{K}_j, j = 1, 2\}$. Using now Proposition 1.55 for the pair of indices p and $q = 1$, we have

$$|f| * |g|(x) = \|h_x\|_{L^1} \leq C\|h_x\|_{L^p} = C\left(\int_{\mathbb{R}^d} |f(y)|^p |g(x - y)|^p \, dy\right)^{1/p},$$

where the constant C depends only on the diameters of the compact sets $\mathbb{K}_1, \mathbb{K}_2$ and p. Thus, using the last inequality and Tonelli's Theorem 1.6, we get

$$
\begin{aligned}
\||f| * |g|\|_{L^p} &= \left(\int_{\mathbb{R}^d} (|f| * |g|)^p(x) \, dx\right)^{1/p} \\
&\leq C\left(\int_{\mathbb{R}^d} \left(\int_{\mathbb{R}^d} |f(y)|^p |g(x - y)|^p \, dy\right) dx\right)^{1/p} \\
&= C\|f\|_{L^p}\|g\|_{L^p}.
\end{aligned}
$$

$\square$

Propositions 1.55 and 1.56 can be slightly extended [166, Theorem 2.5 and Lemma 2.6]; these statements and proofs are given next.

Proposition 1.57 *Let $\mathbb{K} \subset \mathbb{R}^d$ be a compact set, $\xi_0 \in \mathbb{R}^d$, and $0 < p \leq q \leq \infty$. If $f \in PW^p(\mathbb{K}; \xi_0)$, then there exists a positive constant $C = C(\operatorname{diam} \mathbb{K}, p)$ such that $\|f\|_{L^q} \leq C\|f\|_{L^p}$.*

Proof Since $\widehat{M_{-\xi_0}f} = T_{-\xi_0}\widehat{f}$ and $f \in L^p \Leftrightarrow M_{-\xi_0}f \in L^p$, we see that $f \in PW^p(\mathbb{K}; \xi_0)$ implies that $M_{-\xi_0}f$ satisfies the conditions of Proposition 1.55. This gives the desired result.   □

Proposition 1.58 *Let $\mathbb{K}_1, \mathbb{K}_2 \subset \mathbb{R}^d$ be compact sets, $\xi_1, \xi_2 \in \mathbb{R}^d$, and $0 < p \leq 1$. If $f \in PW^p(\mathbb{K}_1, \xi_1)$ and $g \in PW^p(\mathbb{K}_2, \xi_2)$, then there exists a constant $C = C(\mathbb{K}_1, \mathbb{K}_2, p)$ such that $\||f| * |g|\|_{L^p} \leq C\|f\|_{L^p}\|g\|_{L^p}$.*

Proof Proceeds similar to the proof of Proposition 1.57, but now using the statement of Proposition 1.56.   □

Somewhat similar in spirit to the statements above is the following result, see [227, Theorem 1.4.1].

Proposition 1.59 *Let $\mathbb{K}$ be a compact subset of $\mathbb{R}^d$, $\xi_0 \in \mathbb{R}^d$, $0 < p \leq \infty$, and $\alpha = \alpha(\mathbb{K}) > 0$ sufficiently small. Given $f \in PW^p(\mathbb{K}; \xi_0)$, let $\mathbf{f}_\alpha$ denote the sequence $(f(\alpha k))_{k \in \mathbb{Z}^d}$. Then, $\|f\|_{L^p} \sim \|\mathbf{f}_\alpha\|_{\ell^p}$, that is*

$$\|f\|_{L^p} \lesssim \left(\sum_{k \in \mathbb{Z}^d} |f(\alpha k)|^p \right)^{1/p} \lesssim \|f\|_{L^p}.$$

1.3.1.2 Bessel Potential Spaces and Fourier-Lebesgue Spaces

Definition 1.60 Let $s \geq 0$. The *inhomogeneous Bessel potential spaces* (or L^p-Sobolev spaces) $L_s^p(\mathbb{R}^d)$, $1 \leq p \leq \infty$ are defined via the complete norms

$$\|f\|_{L_s^p} = \|((1 + 4\pi^2|\xi|^2)^{s/2}\widehat{f}(\xi))^\vee\|_{L^p(\mathbb{R}^d)}.$$

Similarly, we define the *homogeneous Sobolev spaces* $\dot{L}_s^p(\mathbb{R}^d)$ via the norm

$$\|f\|_{\dot{L}_s^p} = \|(|\xi|^s\widehat{f}(\xi))^\vee\|_{L^p(\mathbb{R}^d)}.$$

We will write $H^s(\mathbb{R}^d) = L_s^2(\mathbb{R}^d)$ and $\dot{H}^s = \dot{L}_s^2$ for the L^2-Sobolev spaces. Note that by virtue of Plancherel's theorem,

$$f \in H^s \text{ if and only if } \int_{\mathbb{R}^d} (1 + 4\pi^2|\xi|^2)^s|\widehat{f}(\xi)|^2 \, d\xi < \infty.$$

Clearly, $H^0 = \dot{H}^0 = L^2$. We point out also that, strictly speaking, the homogeneous Sobolev spaces are defined only for equivalence classes modulo polynomials which correspond to distributions having their Fourier transform supported at the origin.

The *Sobolev embedding* is an inequality that is ubiquitous in harmonic analysis and partial differential equations. It can be stated for both (homogeneous and inhomogeneous) versions of the Bessel potential spaces L_s^p, and, in particular, for the Sobolev spaces H^s; see, for example, [9].

Proposition 1.61 (Sobolev) *Let $s > 0$ and $1 < p < q < \infty$. If $\frac{s}{d} = \frac{1}{p} - \frac{1}{q}$, then*

$$\|f\|_{L^q(\mathbb{R}^d)} \lesssim \|f\|_{\dot{L}_s^p(\mathbb{R}^d)}.$$

Also, if $\frac{s}{d} \geq \frac{1}{p} - \frac{1}{q}$, then

$$\|f\|_{L^q(\mathbb{R}^d)} \lesssim \|f\|_{L_s^p(\mathbb{R}^d)}.$$

We note that the end-point $q = \infty$ can be included in the inhomogeneous version as long as the inequality of the exponents is strict, that is, $sp > d$. By letting $p = 2$ in Proposition 1.61, we get

$$\|f\|_{L^q(\mathbb{R}^d)} \lesssim \|f\|_{\dot{H}^s(\mathbb{R}^d)}$$

as long as $q > 2$ and $s/d = 1/2 - 1/q$.

Definition 1.62 Let $1 \leq p \leq \infty$. The *Fourier-Lebesgue space* $\mathscr{F}L^p$ is the Banach space of tempered distributions $f \in \mathscr{S}'$ endowed with the norm

$$\|f\|_{\mathscr{F}L^p} = \|\widehat{f}\|_{L^p}.$$

More generally, for $s \geq 0$, we define the spaces $\mathscr{F}L^{s,p}$ via the norm

$$\|f\|_{\mathscr{F}L^{s,p}} = \|\langle \xi \rangle^s \widehat{f}(\xi)\|_{L^p},$$

where we use the "Japanese bracket" notation $\langle \xi \rangle = (1 + 4\pi^2|\xi|^2)^{1/2}$.

The Riemann-Lebesgue Lemma 1.45 and a straightforward calculation show that the following inclusions hold: $\mathscr{F}L^1(\mathbb{R}^d) \subset \mathscr{C}_0(\mathbb{R}^d)$ and, for $s > d/2$, $H^s(\mathbb{R}^d) \subset \mathscr{F}L^1(\mathbb{R}^d)$. We note also that $L_s^p(\mathbb{R}^d)$ (for $s > d/p$), and, in particular, $H^s(\mathbb{R}^d)$ (for $s > d/2$), are multiplication algebras, see [227, Theorem 2.8.3]. More precisely, we have the following statement.

Proposition 1.63 *Let $1 \leq p \leq \infty$ and $s > d/p$. If $f, g \in L_s^p(\mathbb{R}^d)$, then $fg \in L_s^p(\mathbb{R}^d)$ and*

$$\|fg\|_{L_s^p} \lesssim \|f\|_{L_s^p} \|g\|_{L_s^p}.$$

We briefly indicate below the argument which shows that if $s > d/2$, then $H^s(\mathbb{R}^d)$ is a multiplication algebra; a similar argument works for $L_s^p(\mathbb{R}^d)$.

Proof It is immediate to see that for all $s \geq 0$, we have

$$\langle \xi \rangle^s \leq \langle \xi - \eta \rangle^s \langle \eta \rangle^s \text{ and } \langle \xi \rangle^s \lesssim \langle \xi - \eta \rangle^s + \langle \eta \rangle^s.$$

Thus, for $f, g \in H^s$ ($s > d/2$), we can write

$$\|fg\|_{H^s} = \|\langle \cdot \rangle^s \widehat{fg}\|_{L^2} = \|\langle \cdot \rangle^s \widehat{f} * \widehat{g}\|_{L^2}$$

$$= \|\langle \xi \rangle^s \int_\eta \widehat{f}(\xi - \eta) \widehat{g}(\eta) d\eta\|_{L^2(d\xi)}$$

$$\lesssim A + B,$$

where

$$A = \left\| \int_\eta \langle \xi - \eta \rangle^s |\widehat{f}(\xi - \eta)| |\widehat{g}(\eta)| d\eta \right\|_{L^2(d\xi)}, \quad B = \left\| \int_\eta \langle \eta \rangle^s |\widehat{f}(\xi - \eta)| |\widehat{g}(\eta)| d\eta \right\|_{L^2(d\xi)}.$$

We further control the term A by using Young's inequality (Theorem 1.11) and the embedding $H^s \subset \mathscr{F}L^1$ as follows:

$$A = \|(\langle \cdot \rangle^s |\widehat{f}|) * |\widehat{g}|\|_{L^2} \leq \|\langle \cdot \rangle^s |\widehat{f}|\|_{L^2} \|\widehat{g}\|_{L^1}$$

$$= \|f\|_{H^s} \|g\|_{\mathscr{F}L^1} \lesssim \|f\|_{H^s} \|g\|_{H^s}.$$

We control B in a similar way, observing now that $B = \|(\langle \cdot \rangle^s |\widehat{g}|) * |\widehat{f}|\|_{L^2}$. This completes the proof. □

The following interpolation results are known.

Proposition 1.64 *Let* $1 \leq p_1, p_2 \leq \infty$, $s_1, s_2 \geq 0$, $0 \leq \theta \leq 1$. *Define*

$$s = \theta s_1 + (1 - \theta) s_2, \quad \frac{1}{p} = \frac{\theta}{p_1} + \frac{1 - \theta}{p_2}.$$

Then, if $f \in L_{s_1}^{p_1} \cap L_{s_2}^{p_2}$, *we have* $f \in L_s^p$ *and*

$$\|f\|_{L_s^p} \lesssim \|f\|_{L_{s_1}^{p_1}} \|f\|_{L_{s_2}^{p_2}}.$$

Proposition 1.65 (Bernstein) *Let* $1 \leq p \leq 2$ *and let* $k > d/p$ *be an integer. Then*

$$\|f\|_{\mathscr{F}L^1} \lesssim \|f\|_{L^p}^{1 - \frac{d}{2k}} \|f\|_{L_k^p}^{\frac{d}{kp}}.$$

We present now a result known as *Bernstein's multiplier theorem*, see Hörmander [150]; compare also to Proposition 1.55.

Proposition 1.66 (Bernstein) *Let $\mathbb{K} \subset \mathbb{R}^d$ be a compact set and $0 < p \leq \infty$. Let $q > q(p, d) := d\left(\frac{1}{\min(1, p)} - \frac{1}{2}\right)$. If $f \in PW^p(\mathbb{K})$ and $\psi \in H^q$, then there exists a constant $C > 0$ such that $\|\mathscr{F}^{-1}\psi\mathscr{F}f\|_{L^p} \leq C\|\psi\|_{H^q}\|f\|_{L^p}$. Moreover, if $p \geq 1$, the previous multiplier estimate holds under the weaker condition $f \in L^p$.*

To be precise, when we write $\mathscr{F}^{-1}\psi\mathscr{F}f$, we mean $(\psi\widehat{f})^{\vee}$.

Finally, we state a simple result for smooth functions with compact support; see [66].

Proposition 1.67 *Assume that $f \in \mathscr{C}_c^{\infty}(\mathbb{R}^d)$. Then, for every $p > 0$ there exists $N = \lceil\frac{d}{2p}\rceil \in \mathbb{N}$ such that $\|f\|_{\mathscr{F}L^p} \lesssim |\mathrm{supp}\,(f)|\sup_{|\alpha|\leq N}\|\partial^{\alpha}f\|_{L^{\infty}}$.*

Proof The estimate is straightforward:

$$\|f\|_{\mathscr{F}L^p} = \|\widehat{f}\|_{L^p} = \||\widehat{f}|^p\|_{L^1}^{1/p}$$

$$= \|\langle\xi\rangle^{-2Np}(\langle\xi\rangle^{2N}|\widehat{f}(\xi)|)^p\|_{L^1}^{1/p}$$

$$= \left(\int_{\mathbb{R}^d}\langle\xi\rangle^{-2Np}\left|\int_{\mathbb{R}^d}(I - \Delta)^N f(x)e^{-2\pi ix\cdot\xi}\,dx\right|^p d\xi\right)^{1/p}$$

$$\lesssim |\mathrm{supp}\,(f)|\sup_{|\alpha|\leq N}\|\partial^{\alpha}f\|_{L^{\infty}}.$$

$\square$

1.3.1.3 The Fourier Transform of a Distribution

We have seen that the Fourier transform can be extended to L^2 and

$$\langle\widehat{f}, g\rangle_{L^2} = \langle f, g^{\vee}\rangle_{L^2}.$$

It is thus natural to extend the Fourier transform to distributions. More precisely, if $u \in \mathscr{S}'$, we can define $\widehat{u} \in \mathscr{S}'$ by essentially replacing the L^2-inner product with the dual bracket $\langle\cdot, \cdot\rangle_{\mathscr{S}',\mathscr{S}}$; for all $g \in \mathscr{S}$, we let

$$\langle\widehat{u}, g\rangle = \langle u, \widehat{g}\rangle.$$

It is a straightforward calculation to show that $\mathscr{F} : \mathscr{S}' \to \mathscr{S}'$ is an isomorphism and $\mathscr{F}^{-1}u = u^{\vee} = \widehat{\tilde{u}}$ for all $u \in \mathscr{S}'$. All properties listed in Proposition 1.46 now transfer almost verbatim to tempered distributions; the exception is that in the convolution property we now have $\widehat{u * g} = \widehat{u}\widehat{g}$ for all $u \in \mathscr{S}'$ and $g \in \mathscr{S}$.

Example 1.68 We have $\widehat{\delta_0} = 1$. Indeed, for all $f \in \mathscr{S}$, we can write

$$\langle\widehat{\delta_0}, f\rangle = \langle\delta_0, \widehat{f}\rangle = \widehat{f}(0) = \int_{\mathbb{R}^d} f(x)\,dx = \langle 1, f\rangle.$$

Consequently, we also obtain $\widehat{1} = \delta_0$.

Example 1.69 We have $\widehat{\delta_y} = M_y 1$. For $f \in \mathscr{S}$, we compute as above

$$\langle \widehat{\delta_y}, f \rangle = \widehat{f}(y) = \int_{\mathbb{R}^d} f(x) e^{-2\pi i y \cdot x} \, dx = \langle 1, M_{-y} f \rangle = \langle M_y 1, f \rangle.$$

Alternately, $\widehat{\delta_y} = \widehat{T_{-y} \delta_0} = M_y \widehat{\delta_0} = M_y 1$.

Example 1.70 Let $u = \text{p.v.} \frac{1}{x}$ be the distribution defined in Example 1.38. We compute $\widehat{u}$. Note first that $xu = 1$, that is, for all $f \in \mathscr{S}$ we have $\langle xu, f \rangle = \langle u, xf \rangle = \int_{\mathbb{R}} f(x) \, dx = \langle 1, f \rangle$. Therefore, $\widehat{xu} = \widehat{1} = \delta_0$. Thus, by Proposition 1.46, we get that $\widehat{u}' = -2\pi i \delta_0$. By Example 1.41 we conclude that $\widehat{u} = -\pi i \, \text{sign}\, x$.

We wish to end this subsection with a brief discussion regarding the interaction between the "concentration" of a function and that of its Fourier transform. Recalling the Gaussian, we see that φ_1 and $\widehat{\varphi_1}$ are equally spread out (over the whole $\mathbb{R}$). At the other extreme, we can consider the Dirac distribution δ_0 concentrated at the origin, where we go from concentration at a point in the time domain to no concentration at all in the frequency domain. Thus, we expect in general that the more we concentrate a signal in the time domain, the less we can concentrate the signal in the frequency domain. One instance of this phenomenon is given in the following statement.

Theorem 1.71 (Paley-Wiener) *If $f \in L^2(\mathbb{R}) \setminus \{0\}$ and $\widehat{f}$ has compact support, then f is the restriction to $\mathbb{R}$ of an entire function. In particular, f and $\widehat{f}$ cannot have both compact support.*

Another instance of the phenomenon described above is given by the *uncertainty principle*, which makes the concept of instantaneous frequency impossible. Loosely speaking, this principle states that for $f \in L^2(\mathbb{R})$, f and $\widehat{f}$ cannot be both concentrated around the same point in the time-frequency plane.

Theorem 1.72 (Heisenberg's Uncertainty Principle) *For any $f \in L^2(\mathbb{R})$ with $\|f\|_{L^2} = 1$, we have*

$$\min_{x_0 \in \mathbb{R}} \|(x - x_0) f(x)\|_{L^2(\mathbb{R})} \min_{\xi_0 \in \mathbb{R}} \|(\xi - \xi_0) \widehat{f}(\xi)\|_{L^2(\mathbb{R})} \geq \frac{1}{16\pi^2}$$

with equality if and only if $f(x) = e^{2\pi i b(x-a)} e^{-\pi(x-a)^2/c}$ for some $a, b \in \mathbb{R}$ and $c > 0$.

Thus, the joint description of the time and frequency behavior of a signal is by necessity a two-dimensional one. This leads to the short-time Fourier transform of a signal, which we explore next.

1.3.2 The Short-Time Fourier Transform

The *short-time Fourier transform* of a signal f at time x is obtained by localizing f, using a smooth cut-off function ϕ, to a neighborhood of x and then taking the Fourier transform.

Definition 1.73 Let $\phi \in \mathscr{S}(\mathbb{R}^d)$ be a fixed, nonzero *window function* and $u \in \mathscr{S}'(\mathbb{R}^d)$ a tempered distribution. The short-time Fourier transform (STFT) of u with respect to ϕ is defined as

$$V_\phi u(x, \xi) = \langle u, M_\xi T_x \phi \rangle.$$

In general, we can define the STFT as long as the window function ϕ belongs to a space of test functions $\mathscr{X}$ that is invariant under time-frequency shifts and f is in its dual space $\mathscr{X}'$. The STFT can also be defined when *both* u and ϕ are in $\mathscr{S}'$, see [108, Proposition 1.42]. Assume now that the tempered distribution is (identified with) a function $f \in L^2(\mathbb{R}^d)$. Then, using Parseval's identity (see Proposition 1.51), the properties of the Fourier transform (see Proposition 1.46), and the almost commutation property of the time-frequency shifts (1.9) we obtain several equivalent forms of the STFT; we note that the integral version exists only for functions of polynomial growth.

$$V_\phi f(x, \xi) = \int_{\mathbb{R}^d} f(y)\overline{\phi(y - x)}e^{-2\pi i y \cdot \xi} \, dy = \mathscr{F}(f\overline{T_x\phi})(\xi) \tag{1.12}$$

$$= \langle f, M_\xi T_x \phi \rangle_{L^2} = \langle \widehat{f}, \widehat{M_\xi T_x \phi} \rangle_{L^2} \tag{1.13}$$

$$= \langle \widehat{f}, T_\xi M_{-x} \widehat{\phi} \rangle_{L^2} = e^{-2\pi i x \cdot \xi} \langle \widehat{f}, M_{-x} T_\xi \widehat{\phi} \rangle_{L^2} \tag{1.14}$$

$$= \left(\widehat{f} * M_{-x}\overline{\widetilde{\phi}}\right)(\xi) = e^{-2\pi i x \cdot \xi}(f * M_\xi \overline{\widetilde{\phi}})(x). \tag{1.15}$$

Next, we compute the STFT on a few basic examples, in particular, the STFT of the d-dimensional Gaussian function with respect to the d-dimensional Gaussian window.

Example 1.74 Let $f = \varphi_d \in \mathscr{S}(\mathbb{R}^d)$ and $\phi = \varphi_d \in \mathscr{S}(\mathbb{R}^d)$; recall that $\varphi_d(x) = e^{-\pi|x|^2}$. Recall also that, in Example 1.47, we have shown that the d-dimensional Gaussian is left invariant by the Fourier transform. We claim that

$$V_\phi f(x, \xi) = 2^{d/2}e^{-\pi i x \cdot \xi} D_{2^{1/2}}\varphi_{2d}(x, \xi) = 2^{-\frac{d}{2}}e^{-\pi i x \cdot \xi}e^{-\frac{\pi}{2}(|x|^2 + |\xi|^2)}.$$

This follows, as in Example 1.47, from an appropriate completion of squares and properties of the Fourier transform (see again Proposition 1.46). We have

$$V_\phi f(x, \xi) = \langle f, M_\xi T_x \phi \rangle_{L^2} = \int_{\mathbb{R}^d} e^{-\pi |y|^2} e^{-\pi |y-x|^2} e^{-2\pi i y \cdot \xi} \, dy$$

$$= e^{-\pi |x|^2/2} \int_{\mathbb{R}^d} e^{-2\pi |y-\frac{x}{2}|^2} e^{-2\pi i y \cdot \xi} \, dy$$

$$= 2^{-d/2} e^{-\pi |x|^2/2} \mathscr{F}(T_{x/2} D_{2^{-1/2}} \varphi_d)(\xi) = 2^{-d/2} e^{-\pi |x|^2/2} M_{-x/2} \widehat{D_{2^{-1/2}} \varphi_d}(\xi)$$

$$= 2^{-d/2} e^{-\pi |x|^2/2} e^{-\pi i x \cdot \xi} \widehat{\varphi_d}(2^{-1/2} \xi) = 2^{-d/2} e^{-\pi i x \cdot \xi} e^{-\pi |x|^2/2} e^{-\pi |\xi|^2/2}.$$

Example 1.75 Let $f(x) = e^{i\pi x^2}$, $x \in \mathbb{R}$. We proceed similarly as in Example 1.74 and obtain $|V_{\varphi_1} f(x, \xi)| = c e^{-\frac{\pi}{2}(x-\xi)^2}$, where c is some positive constant.

Example 1.76 More generally, let $t \in \mathbb{R}$, $\alpha \geq 0$, and define $\sigma_{t;\alpha}(x) = e^{i\pi t |x|^\alpha}$, $x \in \mathbb{R}^d$. With $\phi(x) = \varphi_d(x) = e^{-\pi |x|^2}$ as in Example 1.74, we get after some straightforward manipulations

$$V_\phi \sigma_{t;2}(x, \xi) = e^{-\pi |x|^2} \int_{\mathbb{R}^d} e^{-\pi(1-it)|y|^2} e^{2\pi y \cdot x} e^{-2\pi i y \cdot \xi} \, dy$$

$$= e^{-\pi |x|^2} (1 - it)^{-d/2} e^{\pi(1-it)|x|^2} T_{tx} M_{-x} \left(e^{-\pi |\xi|^2/(1-it)} \right),$$

where we consider the square root $(1 - it)^{1/2}$ with positive imaginary part. Thus, by taking the absolute values on both sides of this equality, we arrive to

$$|V_{\varphi_d} \sigma_{t;2}(x, \xi)| = (1 + t^2)^{-d/4} e^{-\frac{\pi}{1+t^2} |tx - \xi|^2}.$$

Example 1.77 Consider now $\varphi_{2d}(x, \xi) = e^{-\pi(|x|^2 + |\xi|^2)}$, and for $a, b, c > 0$ define the *generalized Gaussian function*

$$f_{a,b,c}(x, \xi) = e^{-\pi a |x|^2} e^{-\pi b |\xi|^2} e^{2\pi i c x \cdot \xi};$$

that is, $f_{a,b,c}$ is an appropriately modulated version of the tensor product of appropriately dilated d-dimensional Gaussians. Let $\alpha = \alpha(a, b, c) := (a + 1)(b + 1) + c^2$. Then, letting $y = (y_1, y_2)$, $\eta = (\eta_1, \eta_2) \in \mathbb{R}^{2d}$, we have the following identity (see [69, Proposition 2.2]):

$$V_{\varphi_{2d}} f_{a,b,c}(y, \eta) =$$

$$\alpha^{d/2} e^{-\frac{\pi}{\alpha} \left((a(b+1)+c^2)|y_1|^2 + (b(a+1)+c^2)|y_2|^2 + (b+1)|\eta_1|^2 + (a+1)|\eta_2|^2 - 2c(y_1 \cdot \eta_2 + y_2 \cdot \eta_1) \right)}$$

$$\times e^{-\frac{2\pi i}{a+1} \left(y_1 \cdot \eta_1 + \alpha^{-1}(cy_1 - (a+1)\eta_2) \cdot (c\eta_1 + (a+1)y_2) \right)}.$$

Example 1.78 For a fixed nonzero window function $\phi \in \mathscr{S}(\mathbb{R}^d)$ and for all $x, \xi \in \mathbb{R}^d$, we have $V_\phi \delta_0(x, \xi) = \overline{\tilde{\phi}(x)}$. This follows from the elementary calculation

$$V_\phi \delta_0(x, \xi) = \langle \delta_0, M_\xi T_x \phi \rangle = \overline{M_\xi T_x \phi(y)}|_{y=0} = \overline{\phi(-x)}.$$

In particular, if $\phi = \varphi_d$, we get $V_{\varphi_d} \delta_0(x, \xi) = e^{-\pi |x|^2}$.

Example 1.79 Similar to the previous example, we can show that for $f \equiv 1$, we have $V_\phi f(x, \xi) = e^{-2\pi i x \cdot \xi} \overline{\widehat{\phi}(\xi)}$, and, in particular, $V_{\varphi_d} 1(x, \xi) = e^{-2\pi i x \cdot \xi} e^{-\pi |\xi|^2}$.

Perhaps unsurprisingly, in view of the properties of the Fourier transform, the STFT also interacts rather nicely with the operations of translation, modulation, and dilation. Straightforward changes of variables prove the identities in the next lemma; see Proposition 1.46 for a comparison with the properties of the Fourier transform.

Lemma 1.80 *Let* $\phi \in \mathscr{S}(\mathbb{R}^d) \setminus \{0\}$, $f \in L^2(\mathbb{R}^d)$, $z, \eta \in \mathbb{R}^d$, *and* $\epsilon > 0$. *Then*

(a) $V_\phi(T_z f)(x, \xi) = e^{-2\pi i z \cdot \xi} V_{T_{-z}\phi} f(x, \xi)$, *and consequently*
(b) $V_{T_z\phi}(T_z f)(x, \xi) = M_{-z}(V_\phi f)(x, \cdot)(\xi)$;
(c) $V_\phi(M_\eta f)(x, \xi) = V_\phi f(x, \xi - \eta) = T_\eta(V_\phi f)(x, \cdot)(\xi)$;
(d) $V_\phi(D_\epsilon f)(x, \xi) = \epsilon^{-d} V_{D_{\epsilon^{-1}}\phi} f(\epsilon^{-1} x, \epsilon \xi)$, *and consequently*
(e) $V_{D_\epsilon\phi}(D_\epsilon f)(x, \xi) = D_\epsilon((V_\phi f)(\cdot, \epsilon \xi))(x)$;
(f) $V_{D_\epsilon\phi} f(x, \xi) = \epsilon^{-d} V_\phi(D_{\epsilon^{-1}} f)(\epsilon^{-1} x, \epsilon \xi)$.

Proof We only sketch the proof of part (f). All the others are similar simple exercises. With $f \in \mathscr{S}$, we have

$$V_{D_\epsilon\phi} f(x, \xi) = \epsilon^{-d} \int_{\mathbb{R}^d} f(y) \overline{\phi(\epsilon^{-1} y - \epsilon^{-1} x)} e^{-2\pi i y \cdot \xi} \, dy$$

$$= \epsilon^{-d} \int_{\mathbb{R}^d} f(\epsilon y) \overline{\phi(y - \epsilon^{-1} x)} e^{-2\pi i y \cdot (\epsilon \xi)} \epsilon^d dy$$

$$= \epsilon^{-d} V_\phi(D_{\epsilon^{-1}} f)(\epsilon^{-1} x, \epsilon \xi).$$

$\square$

Remark 1.81 When measuring the mixed $L^{p,q}$ norm of the STFT of a function in which the window has been dilated, the dilation constant ϵ plays, as expected, an important role. More precisely, using Lemma 1.80 (f), we obtain

$$\|V_{D_\epsilon\phi} f\|_{L^{p,q}} = \epsilon^{d(1/p - 1/q - 1)} \|V_\phi(D_{\epsilon^{-1}} f)\|_{L^{p,q}}.$$

In particular,

$$\|V_{D_\epsilon\phi} f\|_{L^{1,1}} = \epsilon^{-d} \|V_\phi(D_{\epsilon^{-1}} f)\|_{L^{1,1}}.$$

Besides the identities in Lemma 1.80, the STFT has several other desirable properties.

Proposition 1.82 *Let* $\phi, \phi_1, \phi_2 \in \mathscr{S}(\mathbb{R}^d)$ *be (fixed) nonzero windows,* $f, f_1, f_2 \in L^2(\mathbb{R}^d)$, *and* $u \in \mathscr{S}'(\mathbb{R}^d)$. *We have the following.*

(a) $V_\phi u$ is a continuous function on $\mathbb{R}^{2d}$ which is tempered at infinity;

(b) If $f \in \mathscr{S}(\mathbb{R}^d)$, then $V_\phi f \in \mathscr{S}(\mathbb{R}^{2d})$;

(c) (**The fundamental STFT identity**) $V_\phi f(x, \xi) = e^{-2\pi i x \cdot \xi} V_{\widehat{\phi}} \widehat{f}(\xi, -x)$;

(d) (**The covariance of STFT**) $V_\phi(T_y M_\eta f)(x, \xi) = e^{-2\pi i y \cdot \xi} V_\phi f(x - y, \xi - \eta)$;

(e) (**Parseval's STFT identity**) $\langle V_{\phi_1} f_1, V_{\phi_2} f_2 \rangle_{L^2} = \langle f_1, f_2 \rangle_{L^2} \overline{\langle \phi_1, \phi_2 \rangle}_{L^2}$;

(f) (**Plancherel's STFT identity**) $\|V_\phi f\|_{L^2(\mathbb{R}^{2d})} = \|\phi\|_{L^2(\mathbb{R}^d)} \|f\|_{L^2(\mathbb{R}^d)}$;

(g) (**Inverse STFT formula**) If $\langle \phi_1, \phi_2 \rangle_{L^2} \neq 0$, then

$$f = \frac{1}{\langle \phi_2, \phi_1 \rangle_{L^2}} \int_{\mathbb{R}^d} \int_{\mathbb{R}^d} V_{\phi_1} f(x, \xi) M_\xi T_x \phi_2 \, dx \, d\xi.$$

(h) (**The convolution STFT identity**)$\big(V_{\phi_1} f_1(x, \cdot) * V_{\phi_2} f_2(x, \cdot)\big)(\xi)$
$= V_{\phi_1 \phi_2}(f_1 f_2)(x, \xi)$.

Proof For the first two statements, we refer the reader to [125, Section 11.2]. (c) was proved in (1.14). The covariance property (d) follows from the almost commutation relation of time-frequency shifts (1.9):

$$V_\phi(T_y M_\eta f)(x, \xi) = \langle T_y M_\eta f, M_\xi T_x \phi \rangle_{L^2} = \langle f, M_{-\eta} T_{-y} M_\xi T_x \phi \rangle_{L^2}$$
$$= \langle f, e^{2\pi i y \cdot \xi} M_{\xi - \eta} T_{x - y} \phi \rangle_{L^2} = e^{-2\pi i y \cdot \xi} V_\phi f(x - y, \xi - \eta).$$

Note that, in particular, the covariance property of the STFT and Lemma 1.80 immediately imply also the following **STFT identity of time-frequency shifts**:

$$V_{M_\eta T_y \phi}(M_\eta T_y f)(x, \xi) = e^{2\pi (x \cdot \eta - y \cdot \xi)} V_\phi f(x, \xi).$$

Plancherel's STFT identity follows immediately from Parseval's STFT identity by letting $\phi_1 = \phi_2$ and $f_1 = f_2$ there. To prove (e), we use Parseval's identity for the Fourier transform (see Proposition 1.51) and Fubini's Theorem 1.7:

$$\langle V_{\phi_1} f_1, V_{\phi_2} f_2 \rangle_{L^2(\mathbb{R}^{2d})} = \int_{\mathbb{R}^d} \langle \mathscr{F}(f_1 \overline{T_x \phi_1})(\xi), \mathscr{F}(f_2 \overline{T_x \phi_2})(\xi) \rangle_{L^2(d\xi)} dx$$

$$= \int_{\mathbb{R}^d} \langle (f_1 \overline{T_x \phi_1})(y), (f_2 \overline{T_x \phi_2})(y) \rangle_{L^2(dy)} dx$$

$$= \int_{\mathbb{R}^d} \left(\int_{\mathbb{R}^d} f_1(y) \overline{\phi_1(y - x)} \overline{f_2(y)} \phi_2(y - x) \, dy \right) dx$$

$$= \int_{\mathbb{R}^d} f_1(y) \overline{f_2(y)} \left(\overline{\phi_1(y - x)} \phi_2(y - x) \, dx \right) dy$$

$$= \int_{\mathbb{R}^d} f_1(y) \overline{f_2(y)} \, dy \int_{\mathbb{R}^d} \overline{\phi_1(y - x)} \phi_2(y - x) \, dx$$

$$= \langle f_1, f_2 \rangle_{L^2} \overline{\langle \phi_1, \phi_2 \rangle}_{L^2}.$$

To prove the inverse STFT formula (g), we use again Parseval's STFT identity and observe that, for all $g \in L^2(\mathbb{R}^d)$, we have

$$\left\langle \frac{1}{\langle \phi_2, \phi_1 \rangle_{L^2}} \int_{\mathbb{R}^d} \int_{\mathbb{R}^d} V_{\phi_1} f(x, \xi) M_\xi T_x \phi_2 \, dx \, d\xi, g \right\rangle_{L^2} = \frac{1}{\langle \phi_2, \phi_1 \rangle_{L^2}} \langle V_{\phi_1} f, V_{\phi_2} g \rangle_{L^2}$$

$$= \frac{1}{\langle \phi_2, \phi_1 \rangle_{L^2}} \langle \phi_1, \phi_2 \rangle_{L^2} \langle f, g \rangle_{L^2} = \langle f, g \rangle_{L^2}.$$

Finally, let us prove the convolution identity (h). By using the definition of the STFT (1.12) and the convolution property of the Fourier transform (see Proposition 1.46), we have

$$\begin{aligned}
\left(V_{\phi_1} f_1(x, \cdot) * V_{\phi_2} f_2(x, \cdot) \right)(\xi) &= \left(\mathcal{F}(f_1 \overline{T_x \phi_1}) * \mathcal{F}(f_2 \overline{T_x \phi_2}) \right)(\xi) \\
&= \mathcal{F}(f_1 \overline{T_x \phi_1} f_2 \overline{T_x \phi_2})(\xi) \\
&= \mathcal{F}\left(f_1 f_2 \overline{T_x (\phi_1 \phi_2)} \right)(\xi) \\
&= V_{\phi_1 \phi_2}(f_1 f_2)(x, \xi).
\end{aligned}$$

$\square$

We wish to conclude this section by pointing out that *uncertainty principles* exist and apply also to the STFT. We only list here two beautiful inequalities due to Lieb [177] and Gröchenig [125].

Theorem 1.83 (Lieb's Uncertainty Principle) *Let $\phi \in \mathscr{S}(\mathbb{R}^d)$ and $f \in L^2(\mathbb{R}^d)$ with $\|\phi\|_{L^2} = \|f\|_{L^2} = 1$, and $p \in [2, \infty)$. We have*

$$\|V_\phi f\|_{L^p(\mathbb{R}^{2d})} \leq \left(\frac{2}{p} \right)^{d/p}.$$

The proof of this theorem is a nice application of the Fubini theorem (see Theorem 1.7), Hausdorff-Young inequality (see Proposition 1.52), and Young's inequality (see Theorem 1.11); see [125, Theorem 3.3.2] for the argument. Using Lieb's uncertainty principle, Gröchenig proved in [125, Theorem 3.3.3] the following.

Theorem 1.84 (Gröchenig's Uncertainty Principle) *Let $\phi \in \mathscr{S}(\mathbb{R}^d)$ and $f \in L^2(\mathbb{R}^d)$ with $\|\phi\|_{L^2} = \|f\|_{L^2} = 1$, and $p \in (2, \infty)$. If $\mathbb{S} \subset \mathbb{R}^{2d}$ and $\epsilon \geq 0$ are such that $\|V_\phi f\|_{L^2(\mathbb{S})} \geq 1 - \epsilon$, then*

$$|\mathbb{S}| \geq \sup_{p>2} (1 - \epsilon)^{\frac{2p}{p-2}} \left(\frac{p}{2} \right)^{\frac{2d}{p-2}} \geq 2^d (1 - \epsilon)^4.$$

Note that, in particular, assuming the conditions in the last theorem, we have an absolute lower bound on the support of the STFT:

$$|\text{supp}(V_\phi f)| \geq 2^d.$$

Chapter 2
Modulation Spaces

The (weighted) modulation spaces were first introduced by Feichtinger in the early 1980s who proposed a theory that parallels that of the better known Besov spaces. Though we only consider the modulation spaces on $\mathbb{R}^d$, Feichtinger's original paper [94] was in the general setting of locally compact abelian (LCA, for short) groups. In addition to defining these spaces, Feichtinger also established some foundational results including duality, embedding, and trace theorems. At the center of the definition of modulation spaces is the brilliant idea of imposing natural norm conditions on the convolution $M_\xi \phi * f$ of a modulated window function $\phi \in \mathscr{S}(\mathbb{R}^d)$ and a tempered distribution $f \in \mathscr{S}'(\mathbb{R}^d)$, where $\xi \in \mathbb{R}^d$. By observing that this convolution is just the STFT of f with respect to ϕ it becomes convenient to impose appropriate decay and integrability conditions on the STFT, which is exactly what we will do in this chapter to define the (unweighted) modulation spaces. In a way, this definition of the modulation spaces given in Sect. 2.1 arises from the theory of co-orbit spaces of Feichtinger and Gröchenig [99, 100]. In his fascinating historical account and background for the becoming of the modulation spaces [96], Feichtinger writes that "such norms quantifiably describe properties of functions which are at least as relevant as those described by the standard L^p-norms," and he continues, "any measure preserving transformation acts isometrically on $L^p(\mathbb{R}^d)$, yet typically destroys smoothness and decay properties. In contrast, a modulation space norm would distinguish between smooth rearrangements and non-smooth ones of a given function." Most of the basic properties we discuss in Sect. 2.1 are essentially collected from [125, Chapter 11] and can be viewed as exercises in Fourier and functional analysis. In addition, we place a big emphasis on the dilation property of modulation spaces, see Sect. 2.2, as well as the construction of examples of functions and distributions in various modulation spaces. We shall only consider here modulation spaces with parameters $1 \leq p, q \leq \infty$. The case $0 < p, q < 1$ will be considered in Chap. 3, where we will also come back to Feichtinger's original definition and formally establish its equivalence to the definition given here, while the weighted modulation spaces will be introduced in Chap. 5. It is worth pointing

© Springer Science+Business Media, LLC, part of Springer Nature 2020
Á. Bényi, K. A. Okoudjou, *Modulation Spaces*, Applied and Numerical
Harmonic Analysis, https://doi.org/10.1007/978-1-0716-0332-1_2

out that Rauhut [198] extended the theory of co-orbit spaces and used this extension
to define the modulation spaces for all $0 < p, q \leq \infty$.

While Feichtinger's motivations in defining the modulation spaces are rooted in
the theory of harmonic analysis on LCA groups, they also foreshadow some of the
modern applications in applied and computational harmonic analysis [103, 104].
Moreover, as we will see in Chaps. 4 and 7, the modulation spaces appear naturally
in the study of pseudodifferential operators and certain partial differential equations
(PDEs). To illuminate the attractiveness of these spaces, let us only recall for now
Sjöstrand introduction of a space of tempered distributions, the so-called Sjöstrand
class, in relation to the boundedness of pseudodifferential operators on $L^2(\mathbb{R}^d)$
[206] which turned out to be one of Feichtinger's modulation spaces, and the
rediscovery of modulation spaces in the context of nonlinear evolution PDEs [136].
We will elaborate more on both of these applications and provide further context to
their relevance in the appropriate chapters.

2.1   The Co-orbit Definition and Basic Properties

Definition 2.1 Let $1 \leq p, q \leq \infty$, and $\phi \in \mathscr{S}(\mathbb{R}^d) \setminus \{0\}$. The modulation space
$\mathscr{M}^{p,q} = \mathscr{M}^{p,q}(\mathbb{R}^d)$ is the set of all tempered distribution $f \in \mathscr{S}'(\mathbb{R}^d)$ for which
$V_\phi f \in L^{p,q}(\mathbb{R}^{2d})$, that is

$$\|f\|_{\mathscr{M}^{p,q}} := \left(\int_{\mathbb{R}^d} \left(\int_{\mathbb{R}^d} |V_\phi f(x, \xi)|^p dx \right)^{q/p} d\xi \right)^{1/q} < \infty. \qquad (2.1)$$

When $p = \infty$ or $q = \infty$, the essential supremum is used.

It is a straightforward exercise to prove that (2.1) defines a norm. As a matter
of notation, it is common to write $\mathscr{M}^p$ instead of $\mathscr{M}^{p,p}$. Let us also notice that
Definition 2.1 makes sense if we allow the indices p, q to belong to the larger set
$(0, \infty]$; in this case, (2.1) would only give, in general, a quasi-norm on $\mathscr{M}^{p,q}$. As
we will see in a subsequent chapter, when the indices are positive sub-unitary reals,
it is more convenient to use some other equivalent norms to discuss these spaces.

The following lemma is proved in [125, Lemma 11.3.3].

Lemma 2.2 *Let* $\phi, \varphi, \gamma \in \mathscr{S}(\mathbb{R}^d)$ *such that* $\langle \gamma, \varphi \rangle \neq 0$ *and* $f \in \mathscr{S}'(\mathbb{R}^d)$. *Then*

$$|V_\phi f(x, \xi)| \leq \frac{1}{|\langle \gamma, \varphi \rangle|} (|V_\varphi f| * |V_\phi \gamma|)(x, \xi)$$

for all $(x, \xi) \in \mathbb{R}^{2d}$.

The inversion formula for the STFT stated for L^2 functions in part (g) of
Proposition 1.82 generalizes to all the modulation spaces. We refer to [125,
Proposition 11.3.2] for details.

Proposition 2.3 *Let $1 \leq p, q \leq \infty$. Assume that $\phi, \varphi, \gamma \in \mathscr{S}(\mathbb{R}^d)$. For $F \in \mathscr{S}'(\mathbb{R}^{2d})$, define*

$$V_\phi^* F = \iint_{\mathbb{R}^{2d}} F(x, \xi) M_\xi T_x \phi \, dx \, d\xi,$$

where the integral is interpreted in the weak sense: if $f \in \mathscr{S}(\mathbb{R}^d)$,

$$\langle V_\phi^* F, f \rangle = \iint_{\mathbb{R}^{2d}} F(x, \xi) \overline{V_\phi f(x, \xi)} \, dx \, d\xi = \langle F, V_\phi f \rangle.$$

Then,

(a) *V_ϕ^* maps $L^{p,q}(\mathbb{R}^{2d})$ into $\mathscr{M}^{p,q}(\mathbb{R}^d)$ and satisfies*

$$\|V_\phi^* F\|_{\mathscr{M}^{p,q}} \leq C \|V_\varphi \phi\|_{L^{1,1}} \|F\|_{L^{p,q}}.$$

(b) *In particular, if $F = V_\gamma f$ with $f \in \mathscr{M}^{p,q}(\mathbb{R}^d)$, then the inversion formula*

$$f = \frac{1}{\langle \gamma, \phi \rangle} \iint_{\mathbb{R}^{2d}} V_\phi f(x, \xi) M_\xi T_x \gamma \, d\xi \, dx$$

holds in $\mathscr{M}^{p,q}(\mathbb{R}^d)$. That is, $I_{\mathscr{M}^{p,q}} = \langle \gamma, \phi \rangle^{-1} V_\gamma^* V_\phi$.

(c) *The definition of the modulation space $\mathscr{M}^{p,q}$ is independent of the window $\phi \in \mathscr{S}(\mathbb{R}^d)$. In particular, there exist constants $C_1, C_2 > 0$ such that for all $f \in \mathscr{M}^{p,q}$,*

$$C_1 \|V_\phi f\|_{L^{p,q}} \leq \|V_\varphi f\|_{L^{p,q}} \leq C_2 \|V_\phi f\|_{L^{p,q}}. \tag{2.2}$$

Proof

(a) We first show that if $F \in L^{p,q}(\mathbb{R}^{2d})$, then $V_\phi^* F$ is a well-defined element of $\mathscr{S}'(\mathbb{R}^d)$. Let $\varphi \in \mathscr{S}(\mathbb{R}^d)$. Then, since $V_\phi \varphi \in \mathscr{S}(\mathbb{R}^{2d})$, for some positive constant C and some positive integers N, M we have:

$$|\langle V_\phi^* F, \varphi \rangle| = |\langle F, V_\phi \varphi \rangle|$$

$$\leq \|F\|_{L^{p,q}} \|V_\phi \varphi\|_{L^{p',q'}}$$

$$\leq C \|F\|_{L^{p,q}} \sup_{x, \xi \in \mathbb{R}^d} (1 + |x|^2)^{N/2} (1 + |\xi|^2)^{M/2} |V_\phi \varphi(x, \xi)|.$$

By [125, Corollary 11.2.6] (see also Sect. 1.2.2) the family

$$\left\{ \sup_{x, \xi \in \mathbb{R}^d} (1 + |x|^2)^{N/2} (1 + |\xi|^2)^{M/2} |V_\phi \varphi(x, \xi)| : M, N \geq 0 \right\}$$

forms a collection of equivalent seminorms for $\mathscr{S}(\mathbb{R}^d)$. Consequently, $V_\phi^* F \in \mathscr{S}'(\mathbb{R}^d)$.

Next, we show that $V_\phi^* F \in \mathscr{M}^{p,q}(\mathbb{R}^d)$ by proving that $V_\varphi(V_\phi^* F) \in L^{p,q}(\mathbb{R}^{2d})$.

$$|V_\varphi(V_\phi^* F)(x,\xi)| = |\langle F, V_\phi(M_\xi T_x \varphi)\rangle|$$

$$\leq \iint |F(y,\omega)||V_\phi\varphi(y-x,\omega-\xi)|dy\,d\omega$$

$$= |F| * |\widetilde{V_\phi\varphi}|(x,\xi),$$

where $\widetilde{V_\phi\varphi}(x,\xi) = \overline{V_\phi\varphi(-x,-\xi)}$. Consequently,

$$\|V_\varphi(V_\phi^* F)\|_{L^{p,q}} \leq \||F| * |\widetilde{V_\phi\varphi}|\|_{L^{p,q}}$$

$$\leq \|F\|_{L^{p,q}}\|V_\phi\varphi\|_{L^{1,1}},$$

which proves that $V_\varphi(V_\phi^* F) \in L^{p,q}(\mathbb{R}^{2d})$. Therefore, $V_\phi^* F \in \mathscr{M}^{p,q}(\mathbb{R}^d)$ with

$$\|V_\phi^* F\|_{\mathscr{M}^{p,q}} \lesssim \|V_\varphi\phi\|_{L^{1,1}}\|F\|_{L^{p,q}}.$$

(b) If $f \in \mathscr{M}^{p,q}(\mathbb{R}^d)$, then $F = V_\gamma f \in L^{p,q}(\mathbb{R}^{2d})$, and by (a),

$$g = C^{-1}\iint_{\mathbb{R}^{2d}} V_\gamma f(x,\xi)M_\xi T_x\phi\,dx\,d\xi \in \mathscr{M}^{p,q}(\mathbb{R}^d),$$

where $C = \langle\phi,\varphi\rangle$. Now, for any $h \in \mathscr{S}(\mathbb{R}^d)$, it can be shown [125, Corollary 11.2.7] that

$$h = C^{-1}\iint_{\mathbb{R}^{2d}} V_\gamma h(x,\xi)M_\xi T_x\phi\,dx\,d\xi.$$

This implies that $g = f$ as elements of $\mathscr{S}'(\mathbb{R}^d)$, and the conclusion of this part follows.

(c) Using Lemma 2.2, we have with $C = |\langle\phi,\varphi\rangle|^{-1}$,

$$|V_\phi f(x,\xi)| \leq C|V_\varphi\phi| * |V_\varphi f|(x,\xi).$$

Because $V_\varphi\phi \in \mathscr{S}(\mathbb{R}^{2d})$ and Young's inequality (see Theorem 1.11) we can write

$$\|V_\phi f\|_{L^{p,q}} \leq C\|V_\varphi\phi\|_{L^{1,1}}\|V_\varphi f\|_{L^{p,q}}.$$

The proof is complete.

$\square$

The last part of the previous result establishes that the norm of a distribution in a modulation space is independent of the choice of the particular window used to define the STFT. Moreover, the result quantifies how the norm varies using different windows. We now prove that the modulation spaces are Banach spaces, see [125, Theorem 11.3.5].

Theorem 2.4 *The modulation spaces* $\mathcal{M}^{p,q}$, *with* $1 \leq p, q \leq \infty$, *are Banach spaces.*

Proof Let $\phi \in \mathcal{S}(\mathbb{R}^d)$, and $\{f_n\}_{n=1}^{\infty} \subset \mathcal{M}^{p,q}$ be a Cauchy sequence. Then $\{V_\phi f_n\}_{n=1}^{\infty} \subset L^{p,q}(\mathbb{R}^{2d})$ is a Cauchy sequence. Hence, there exists $F \in L^{p,q}(\mathbb{R}^{2d})$ such that $\lim_{n \to \infty} V_\phi f_n = F$. Using, Proposition 2.3, we define

$$f = \|\phi\|_{L^2}^{-2} \iint_{\mathbb{R}^{2d}} F(x, \xi) M_\xi T_x \phi \, dx \, d\xi.$$

Then, $f \in \mathcal{M}^{p,q}$, and $\lim_{n \to \infty} \|f_n - f\|_{\mathcal{M}^{p,q}} = 0$. $\square$

2.2   Further Properties of the Modulation Spaces

In this section we collect and prove a few results about the modulation spaces. In particular, Theorem 2.5 establishes that the Schwartz class $\mathcal{S}$ is dense in $\mathcal{M}^{p,q}$ when $1 \leq p, q < \infty$. In addition, in Theorem 2.6 we identify the dual of the modulation space $\mathcal{M}^{p,q}$ as the space $\mathcal{M}^{p',q'}$ where $\frac{1}{p} + \frac{1}{p'} = \frac{1}{q} + \frac{1}{q'} = 1$ with $p, q < \infty$. Even in the case where $p = \infty$ or $q = \infty$ one can still prove a similar result by using the notion of Köthe dual [6]. Finally, in Theorem 2.8 it is established that the modulation spaces obey certain natural embeddings.

Theorem 2.5 *Given any* $1 \leq p, q \leq \infty$, *we have that* $\mathcal{S} \subset \mathcal{M}^{p,q}$. *Moreover, if* $p, q < \infty$, *then* $\mathcal{S}$ *is dense in* $\mathcal{M}^{p,q}$.

Proof Let $\phi, f \in \mathcal{S}$ and $\epsilon > 0$. By Proposition 1.82, we know that $V_\phi f \in \mathcal{S}(\mathbb{R}^{2d}) \subset L^{p,q}(\mathbb{R}^{2d})$. Hence, $\mathcal{S} \subset \mathcal{M}^{p,q}$.

Now if $p, q < \infty$, let $f \in \mathcal{M}^{p,q}$ and $\phi \in \mathcal{S}$ such that $\|\phi\|_{L^2} = 1$. For each $n \geq 1$ we define $F_n = V_\phi f \cdot \chi_{\mathbb{K}_n}$ where $\mathbb{K}_n = \{x \in \mathbb{R}^{2d} : |x| \leq n\}$; recall that $\chi_{\mathbb{S}}$ denotes the indicator function of the measurable set $\mathbb{S}$. It is clear that $\lim_{n \to \infty} \|V_\phi f - F_n\|_{L^{p,q}} = 0$ and that

$$f_n(t) = V_\phi^* F_n(t) = \iint_{\mathbb{R}^{2d}} F_n(x, \xi) \, M_\xi T_x \phi(t) \, dx \, d\xi$$

$$= \iint_{K_n} V_\phi f(x, \xi) \, M_\xi T_x \phi(t) \, dx \, d\xi.$$

Because F_n is compactly supported, it is easy to establish that $f_n \in \mathscr{S}(\mathbb{R}^d)$. Using part (b) of Proposition 2.3 we have

$$\|f - f_n\|_{\mathscr{M}^{p,q}} = \|V_\phi^* V_\phi f - V_\phi^* F_n\|_{\mathscr{M}^{p,q}} = \|V_\phi^* (V_\phi f - F_n)\|_{\mathscr{M}^{p,q}} \lesssim \|V_\phi f - F_n\|_{L^{p,q}}.$$

Hence, $\mathscr{S}$ is dense in $\mathscr{M}^{p,q}$ when $p, q < \infty$. $\qquad\qquad\qquad\qquad\qquad\qquad\square$

The next result identifies the dual space of $\mathscr{M}^{p,q}$ with another modulation space when $1 \leq p, q < \infty$. A generalization of this result for $0 < p, q < 1$ will be stated in Proposition 3.10.

Theorem 2.6 *If $1 \leq p, q < \infty$, then the dual of $\mathscr{M}^{p,q}$ is $(\mathscr{M}^{p,q})' = \mathscr{M}^{p',q'}$ under the duality*

$$\langle f, h \rangle = \iint_{\mathbb{R}^{2d}} V_\phi f(x, \xi) \overline{V_\phi h(x, \xi)} dx d\xi$$

for $f \in \mathscr{M}^{p,q}$ and $h \in \mathscr{M}^{p',q'}$.

Proof Let $1 \leq p, q < \infty$, and $\phi \in \mathscr{S}$. Given $f \in \mathscr{M}^{p',q'}$, and $g \in \mathscr{M}^{p,q}$, we have, for some positive constant C that only depends on ϕ and d, that

$$|\langle f, g \rangle| \leq \iint_{\mathbb{R}^{2d}} |V_\phi f(x, \xi)| |V_\phi g(x, \xi)| dx d\xi$$

$$\leq \|V_\phi f\|_{L^{p',q'}} \|V_\phi g\|_{L^{p,q}}$$

$$\leq C\|f\|_{\mathscr{M}^{p',q'}} \|g\|_{\mathscr{M}^{p,q}}.$$

Consequently, $f \in (\mathscr{M}^{p,q})'$, and so

$$\mathscr{M}^{p',q'} \subset (\mathscr{M}^{p,q})'.$$

Conversely, let $f \in (\mathscr{M}^{p,q})'$, and let $\phi \in \mathscr{S}$. For any $g \in L^{p,q}$, we have

$$|\langle V_\phi f, g \rangle| = |\langle f, V_\phi^* g \rangle| \lesssim \|V_\phi^* g\|_{\mathscr{M}^{p,q}} \lesssim \|g\|_{L^{p,q}},$$

where we have used the fact that if $g \in L^{p,q}$, $V_\phi^* g \in \mathscr{M}^{p,q}$ with appropriate norm estimates. From the duality of the mixed-normed Lebesgue spaces, we conclude that $V_\phi f \in L^{p',q'}$ and

$$\|V_\phi f\|_{L^{p',q'}} \lesssim 1.$$

Consequently, $f \in \mathscr{M}^{p',q'}$, which concludes the proof of the theorem. $\qquad\qquad\square$

Remark 2.7 The modulation space $\mathscr{M}^{p,q}$ is an example of Banach function space at least when $p, q < \infty$. In particular, we can define its *associate space* $(\mathscr{M}^{p,q})^*$ as the space of all (measurable) g such that

$$\|g\|_{(\mathscr{M}^{p,q})^*} = \sup \left\{ \left| \int_{\mathbb{R}^d} f(x) g(x) dx \right| : f \in \mathscr{M}^{p,q}, \|f\|_{\mathscr{M}^{p,q}} \leq 1 \right\} < \infty.$$

It is not difficult to show that, when $p, q < \infty$, the associate space of $\mathscr{M}^{p,q}$ is just the dual space $\mathscr{M}^{p',q'}$ where $1/p + 1/q = 1$. Furthermore, even when $p = \infty$ or $q = \infty$ one can show that:

$$\begin{cases} (\mathscr{M}^{p,\infty})^* = \mathscr{M}^{p',1} \\ (\mathscr{M}^{\infty,q})^* = \mathscr{M}^{1,q'} \end{cases}$$

In particular, $(\mathscr{M}^{\infty})^* = \mathscr{M}^1$. We refer to [6] for details about the notion of associate space of a given Banach function space.

The modulation spaces also obey a very simple inclusion relation which is stated in the next result.

Theorem 2.8 *Let $1 \leq p_1 \leq p_2 \leq \infty$, and $1 \leq q_1 \leq q_2 \leq \infty$ be given. Then,*

$$\mathscr{M}^{p_1,q_1} \subset \mathscr{M}^{p_2,q_2}.$$

In particular, for $1 \leq p, q \leq \infty$, we have

$$\mathscr{M}^{1,1} \subset \mathscr{M}^{p,q} \subset \mathscr{M}^{\infty,\infty}.$$

Proof We first assume that $1 < p_1 < p_2 < \infty$, and $1 < q_1 < q_2 < \infty$, and choose $\phi, \varphi \in \mathscr{S}$. Under these assumptions, there are $1 < r, s < \infty$ such that $\frac{1}{p_1} + \frac{1}{r} = 1 + \frac{1}{p_2}$, and similarly, $\frac{1}{q_1} + \frac{1}{s} = 1 + \frac{1}{q_2}$.

Given $f \in \mathscr{M}^{p_1,q_1}$, by Lemma 2.2 we have

$$|V_\phi f(x,\xi)| \lesssim (|V_\varphi \phi| * |V_\varphi f|)(x,\xi).$$

Using Young's inequality, we have

$$\|f\|_{\mathscr{M}^{p_2,q_2}} \lesssim \|V_\phi f\|_{L^{p_2,q_2}}$$

$$\lesssim \left(\int_{\mathbb{R}^d} \left(\int_{\mathbb{R}^d} |V_\phi f(x,\xi)|^{p_2} dx \right)^{q_2/p_2} d\xi \right)^{1/q_2}$$

$$\lesssim \left(\int_{\mathbb{R}^d} \left(\int_{\mathbb{R}^d} [|V_\varphi \phi| * |V_\varphi f|(x,\xi)]^{p_2} dx \right)^{q_2/p_2} d\xi \right)^{1/q_2}$$

$$\lesssim \|V_\varphi \phi\|_{L^{r,s}} \|V_\varphi f\|_{L^{p_1,q_1}}$$

$$\lesssim \|\phi\|_{\mathscr{M}^{r,s}} \|f\|_{\mathscr{M}^{p_1,q_1}}.$$

When $p_1 = 1$ or $q_1 = 1$, then we may take $r = p_2$, and $s = q_2$ in the above arguments.

When $p_2 = 1$ or $q_2 = 1$, then $p_1 = q_1 = 1$ and there is nothing to prove.

When $p_2 = \infty$ or $q_2 = \infty$, we may take $r = p_1'$, and $s = q_1'$. $\square$

In the next section we shall give a few instructive examples of functions and distributions in various modulation spaces. We notice immediately, however, that the space of square integrable functions is an example of a modulation space. Indeed, an application of Plancherel's identity for the STFT, see for example Proposition 1.82, implies that $\mathcal{M}^{2,2} = L^2$.

We recall that for $1 \leq p \leq \infty$, $f \in \mathscr{F}L^p(\mathbb{R}^d)$ if and only if there exists $g \in L^p$ with $f = \widehat{g}$ and

$$\|f\|_{\mathscr{F}L^p} = \|g\|_{L^p}.$$

When restricted to compact sets, the modulation spaces take a very simple form. Let $\mathcal{M}^{p,q}_{\mathrm{comp}}(\mathbb{R}^d)$ denote the subspace of $\mathcal{M}^{p,q}(\mathbb{R}^d)$ consisting of compactly supported functions, and by $\mathcal{M}^{p,q}_{\mathrm{loc}}(\mathbb{R}^d)$ the space of functions that are locally in $\mathcal{M}^{p,q}(\mathbb{R}^d)$. More precisely, $u \in \mathcal{M}^{p,q}_{\mathrm{loc}}(\mathbb{R}^d)$ means that for each $g \in \mathscr{C}_c^\infty(\mathbb{R}^d)$ with $\mathrm{supp}\,(g) \subset \mathbb{K}$ where $\mathbb{K}$ is a compact subset of $\mathbb{R}^d$, we have $u_{\mathbb{K}} = g\,u \in \mathcal{M}^{p,q}$, that is, $u_{\mathbb{K}} \in \mathcal{M}^{p,q}_{\mathrm{comp}}$. The subspaces $(\mathscr{F}L^q)_{\mathrm{comp}}(\mathbb{R}^d)$ and $(\mathscr{F}L^q)_{\mathrm{loc}}(\mathbb{R}^d)$ are defined similarly.

The next result, proved by Kobayashi [166], identifies the compactly supported functions in the modulation space $\mathcal{M}^{p,q}$ to the compactly supported functions in the Fourier Lebesgue space $\mathscr{F}L^q$. We note that some special cases of the result have been established earlier; for instance, the case $p = \infty, q = 1$ was proved in [35, Theorem 5.1], while the case $1 \leq p = q < \infty$ was settled in [95]. See also [8, Proposition B.1], [194], and [203] for similar results.

Proposition 2.9 *Let* $1 \leq p, q \leq \infty$. *Then the following statements hold:*

(a) $\mathcal{M}^{p,q}_{\mathrm{comp}}(\mathbb{R}^d) = (\mathscr{F}L^q)_{\mathrm{comp}}(\mathbb{R}^d)$.
(b) $\mathcal{M}^{p,q}_{\mathrm{loc}}(\mathbb{R}^d) = (\mathscr{F}L^q)_{\mathrm{loc}}(\mathbb{R}^d)$.

Proof

(a) Let $r > 0$ be given and let $f \in \mathscr{F}L^q(\mathbb{R}^d)$ such that $\mathrm{supp}\,(f) \subset \mathbb{B}(0, r)$. Let $\phi \in \mathscr{C}_c^\infty(\mathbb{R}^d)$ with $\mathrm{supp}\,(\phi) \subset \mathbb{B}(0, r)$. Then, for each $\xi \in \mathbb{R}^d$, $V_\phi f(\cdot, \xi)$ is supported in $\mathbb{B}(0, 2r)$. Thus, using the fact that $|V_\phi f(x, \xi)| = |V_{\widehat{\phi}}\widehat{f}(\xi, -x)| = |\mathscr{F}^{-1}(\widehat{f} \cdot T_\xi \overline{\widehat{\phi}})(x)|$, we have the following estimates:

$$\|V_\phi f(\cdot, \xi)\|_{L^p} \leq |\mathbb{B}(0, 2r)|^{1/p}\|V_\phi f(\cdot, \xi)\|_{L^\infty}$$

$$= |\mathbb{B}(0, 2r)|^{1/p}\|\mathscr{F}^{-1}(\widehat{f} \cdot T_\xi \overline{\widehat{\phi}})\|_{L^\infty}$$

$$\leq |\mathbb{B}(0, 2r)|^{1/p}\|\widehat{f} \cdot T_\xi \overline{\widehat{\phi}}\|_{L^1}$$

$$\leq |\mathbb{B}(0, 2r)|^{1/p}(|\widehat{f}| * |\widehat{\phi}|)(\xi).$$

Consequently, $\|V_\phi f\|_{L^{p,q}(\mathbb{R}^{2d})} \leq |\mathbb{B}(0,2r)|^{1/p} \|\widehat{f}\|_{L^q(\mathbb{R}^d)} \|\widehat{\phi}\|_{L^1(\mathbb{R}^d)}$, that is

$$\|f\|_{\mathscr{M}^{p,q}(\mathbb{R}^d)|_{\mathbb{B}(0,r)}} \leq C \|f\|_{\mathscr{F}L^q(\mathbb{R}^d)|_{\mathbb{B}(0,r)}},$$

for some positive constant $C = C(r, , p, q, d)$ depending only on $r, , p, q, d$, which shows

$$\mathscr{F}L^q(\mathbb{R}^d)|_{\mathbb{B}(0,r)} \subset \mathscr{M}^{p,q}(\mathbb{R}^d)|_{\mathbb{B}(0,r)}.$$

For the converse, let $r > 0$ be given and $f \in \mathscr{M}^{p,q}(\mathbb{R}^d)$ such that supp $(f) \subset \mathbb{B}(0, r)$. Let $\phi \in \mathscr{C}_c^\infty(\mathbb{R}^d)$ such that $\phi \equiv 1$ on $\mathbb{B}(0, 2r)$. It is clear that for all $x \in \mathbb{B}(0, r)$ and for all $t \in \mathbb{B}(0, r)$, $\phi(t - x) = 1$. Thus, for all $\xi \in \mathbb{R}^d$ and for $x \in \mathbb{B}(0, r)$,

$$\chi_{\mathbb{B}(0,r)}(x)\widehat{f}(\xi) = \chi_{\mathbb{B}(0,r)}(x) \int_{\mathbb{B}(0,r)} f(t) e^{-2\pi i t \cdot \xi} \overline{\phi(t - x)} \, dt = \chi_{\mathbb{B}(0,r)}(x) V_\phi f(x, \xi).$$

Therefore,

$$|\mathbb{B}(0, r)|^{1/p} |\widehat{f}(\xi)| = \|\chi_{\mathbb{B}(0,r)}(\cdot) V_\phi f(\cdot, \xi)\|_{L^p}.$$

Hence, $\|\widehat{f}\|_{L^q} \leq |\mathbb{B}(0, r)|^{-1/p} \|V_\phi f\|_{L^{p,q}}$. From here we get that

$$\|f\|_{\mathscr{F}L^q(\mathbb{R}^d)|_{\mathbb{B}(0,r)}} \leq C(r, p, q, d) \|f\|_{\mathscr{M}^{p,q}(\mathbb{R}^d)|_{\mathbb{B}(0,r)}},$$

which implies that $\mathscr{M}^{p,q}(\mathbb{R}^d)|_{\mathbb{B}(0,r)} \subset \mathscr{F}L^q(\mathbb{R}^d)|_{\mathbb{B}(0,r)}$.

(b) Let $f \in \mathscr{M}_{\mathrm{loc}}^{p,q}(\mathbb{R}^d)$ and choose $g \in \mathscr{C}_c^\infty$ with supp $(g) \subset \mathbb{K}$, where $\mathbb{K}$ is a compact subset of $\mathbb{R}^d$. Then $f_{\mathbb{K}} = fg \in \mathscr{M}_{\mathrm{comp}}^{p,q}(\mathbb{R}^d) = (\mathscr{F}L^q)_{\mathrm{comp}}(\mathbb{R}^d)$ and the result follows from part (a). $\qquad \square$

The next property essentially states that modulation spaces behave as expected under tensor products.

Lemma 2.10 *Let $F(x_1, x_2) = f_1(x_1) f_2(x_2)$, with $f_j \in \mathscr{S}'(\mathbb{R}^d)$ for $j = 1, 2$. Then $F \in \mathscr{M}^{p,q}(\mathbb{R}^{2d})$ if and only if $f_j \in \mathscr{M}^{p,q}(\mathbb{R}^d)$ for $j = 1, 2$.*

Proof Choose the nonzero window function $\Phi(x_1, x_2) = \phi_1(x_1)\phi_2(x_2)$, with $\phi_j \in \mathscr{S}(\mathbb{R}^d)$ for $j = 1, 2$. Trivially, the STFT of F with respect to Φ splits:

$$V_\Phi F((x_1, x_2), (\xi_1, \xi_2)) = V_{\phi_1} f_1(x_1, \xi_1) V_{\phi_2} f_2(x_2, \xi_2).$$

The conclusion follows immediately from this splitting. $\qquad \square$

We now consider the image of the modulation spaces under various other transformations. We need the following definition.

Definition 2.11 Let $\phi \in \mathscr{S}$. For $1 \leq p, q \leq \infty$, the *Wiener amalgam* $W(\mathscr{F}L^p, L^q)$ is the subset of $\mathscr{S}'(\mathbb{R}^d)$ consisting of all f for which

$$\|f\|_{W(\mathscr{F}L^p, L^q)} := \left(\int_{\mathbb{R}^d} \left(\int_{\mathbb{R}^d} |V_\phi f(x, \xi)|^p d\xi \right)^{q/p} dx \right)^{1/q} < \infty, \qquad (2.3)$$

with appropriate modifications when p or q is infinity.

Proposition 2.12 Let $1 \leq p, q \leq \infty$.

(a) $\mathscr{F}(\mathscr{M}^{p,q}) = W(\mathscr{F}L^p, L^q)$.
(b) If A is any $d \times d$ invertible matrix and we define Δ_A through $\Delta_A f(x) = f(Ax)$, then $\Delta_A(\mathscr{M}^{p,q}) = \mathscr{M}^{p,q}$.

Proof Let $\phi \in \mathscr{S}(\mathbb{R}^d)$.

(a) Assume $f \in \mathscr{M}^{p,q}(\mathbb{R}^d)$. Then, since

$$|V_\phi \widehat{f}(x, \xi)| = |V_{\widehat{\phi}} f(\xi, -x)|,$$

it follows that $\mathscr{F}(\mathscr{M}^{p,q}) = W(\mathscr{F}L^p, L^q)$.
(b) The statement follows from the invariance of the Lebesgue measure under the action of invertible matrices and the following simple calculation:

$$|V_\phi(\Delta_A f)(x, \xi)| = |\det A|^d |V_{\Delta_{A^{-1}}\phi} f(Ax, A^{-1}\xi)|.$$

$\square$

Proposition 2.12 (b) is essentially sharp in the following sense. Given a function $F : \mathbb{R}^d \to \mathbb{R}^d$, we denote by F^* the operator acting on complex-valued functions f by

$$F^*(f)(x) = f(F(x)) = (f \circ F)(x).$$

F^* is called a *superposition operator* or *Nemytskij operator*. Note also that Δ_A in Proposition 2.12 is just a special case of F^*. One can view F^* as a "change of variables" and this change of variables might leave a given modulation space invariant. It turns out that if one assumes an appropriate regularity on F, and if some modulation space is invariant under F^*, then F is essentially a "trivial" mapping. This type of result is known to hold for Besov spaces [202]. Since the definitions of modulation spaces and of the Besov spaces (to be explored and completely clarified in Chap. 6) are related, it is perhaps not entirely surprising that the above phenomenon occurs in the realm of modulation spaces as well. More precisely, Okoudjou [194] proved the following.

Proposition 2.13 *Let* $F : \mathbb{R}^d \to \mathbb{R}^d$ *be a* $\mathscr{C}^1$ *function. Assume that* F^* *maps* $\mathscr{M}^{p,q}$ *into itself, that is,* $F^*(\mathscr{M}^{p,q}(\mathbb{R}^d)) \subset \mathscr{M}^{p,q}(\mathbb{R}^d)$ *for some* $1 \leq p, q \leq \infty$, *with* $2 \neq q < \infty$. *Then,* F *is an affine mapping, that is,* $F(x) = Ax + F(0)$ *for some* $d \times d$ *invertible matrix* A.

The proof is based on a result due to Beurling and Helson [27]; see also [176, 205].

Next, we will explore the action of the dilation operator on modulation spaces. As we shall later see, the dilation property plays a crucial role in the understanding of the optimal embedding between Besov, respectively Sobolev spaces, and modulation spaces. It is also worth pointing out that the dilation (or scaling) property of various function spaces is of great importance when considering the so-called critical regularity conditions in various partial differential equations on these function spaces.

When $A = \lambda I$, where $\lambda > 0$ and I is the $d \times d$ identify matrix, sharp estimates for the modulation space norm of $\Delta_A(f) = f^\lambda$ were proved by Sugimoto and Tomita [211]. Here, we denoted

$$f^\lambda(x) = f(\lambda x).$$

Note that, with the notation introduced in Sect. 1.2.4, $f^\lambda = \lambda^{-d} D_{\lambda^{-1}} f$. Before stating precisely the main result from [211], we provide a few heuristics that reveal the contrast between modulation spaces and Lebesgue spaces under dilation.

In the case of the modulation space $\mathscr{M}^{2,2}(\mathbb{R}^d) = L^2(\mathbb{R}^d)$, we have $\|f^\lambda\|_{\mathscr{M}^{2,2}} \sim \lambda^{-d/2} \|f\|_{\mathscr{M}^{2,2}}$. Clearly, for all $p \geq 1$, we get $\|f^\lambda\|_{L^p(\mathbb{R}^d)} = \lambda^{-d/p} \|f\|_{L^p(\mathbb{R}^d)}$. However, the dependence of $\|f^\lambda\|_{\mathscr{M}^{p,q}}$ on λ is more subtle for generic indices p, q.

To begin with, the scaling depends on the choice of the function f. Fix $\lambda \in (0, 1]$ and $1 \leq p \leq 2$. Then, if $f = \varphi_d$, the d-dimensional Gaussian, we have $\|f^\lambda\|_{\mathscr{M}^{p,\infty}} \sim \lambda^{-d/p}$, while if

$$g(x) = \sum_{k \in \mathbb{Z}^d} M_k T_k \phi(x) = \sum_{k \in \mathbb{Z}^d} e^{ik \cdot x} \phi(x - k)$$

for an appropriate function $\phi \in \mathscr{S}$, we get $\|g^\lambda\|_{\mathscr{M}^{p,\infty}} \sim \lambda^{-2d/p}$; see [211, Lemma 2.1 and Lemma 3.10].

Similarly, while the Lebesgue spaces $L^p(\mathbb{R}^d)$ have a "natural index" with respect to dilation, namely $-d/p$, this phenomenon is not true for modulation spaces. Again, choosing $f = \varphi_d$, one computes that $\|f^\lambda\|_{\mathscr{M}^{p,\infty}} \sim \lambda^{-d}$ for $\lambda \geq 1$, whereas above we have seen $\|f^\lambda\|_{\mathscr{M}^{p,\infty}} \sim \lambda^{-d/p}$ for $\lambda \leq 1$.

In order to state the result in [211] we need to introduce some further notations. Recall first that, for $1 \leq p \leq \infty$, p' denotes the conjugate exponent of p, that is, $1/p + 1/p' = 1$. The following subsets of pairs $(1/p, 1/q) \in [0, 1] \times [0, 1]$ will play a crucial role:

$$\mathbb{I}_1 : \max(1/p, 1/p') \le 1/q, \qquad \mathbb{I}_1^* : \min(1/p, 1/p') \ge 1/q,$$

$$\mathbb{I}_2 : \max(1/q, 1/2) \le 1/p', \qquad \mathbb{I}_2^* : \min(1/q, 1/2) \ge 1/p',$$

$$\mathbb{I}_3 : \max(1/q, 1/2) \le 1/p, \qquad \mathbb{I}_3^* : \min(1/q, 1/2) \ge 1/p.$$

These sets are displayed in Fig. 2.1.

We define now the following two indices:

$$\mu_1(p,q) = \begin{cases} -1/p & \text{if} \quad (1/p, 1/q) \in \mathbb{I}_1^*, \\ 1/q - 1 & \text{if} \quad (1/p, 1/q) \in \mathbb{I}_2^*, \\ -2/p + 1/q & \text{if} \quad (1/p, 1/q) \in \mathbb{I}_3^*, \end{cases} \tag{2.4}$$

and

$$\mu_2(p,q) = \begin{cases} -1/p & \text{if} \quad (1/p, 1/q) \in \mathbb{I}_1, \\ 1/q - 1 & \text{if} \quad (1/p, 1/q) \in \mathbb{I}_2, \\ -2/p + 1/q & \text{if} \quad (1/p, 1/q) \in \mathbb{I}_3. \end{cases} \tag{2.5}$$

These indices are just the "correct" modifications of the indices θ_1, θ_2 defined by Toft in [220]: for $j = 1, 2$, $\theta_j(p,q) = \mu_j(p,q) + 1/p$. We will return to these indices in Chap. 6.

We are ready to state the dilation property of modulation spaces, see [211, Theorem 1.1].

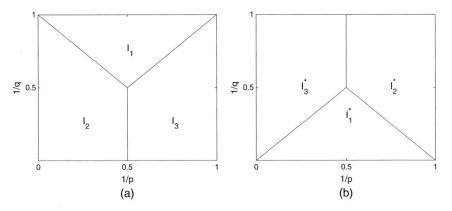

Fig. 2.1 The index sets. (a) $0 < \lambda \le 1$. (b) $\lambda \ge 1$

Theorem 2.14 *Let $1 \leq p, q \leq \infty$. Then the following are true:*

(1) There exists a constant $C > 0$ such that

$$C^{-1}\lambda^{d\mu_2(p,q)}\|f\|_{\mathscr{M}^{p,q}} \leq \|f^\lambda\|_{\mathscr{M}^{p,q}} \leq C\lambda^{d\mu_1(p,q)}\|f\|_{\mathscr{M}^{p,q}}$$

for all $f \in \mathscr{M}^{p,q}$ and all $\lambda \geq 1$. Conversely, if there exist constants $C > 0$ and $\alpha, \beta \in \mathbb{R}$ such that

$$C^{-1}\lambda^\beta\|f\|_{\mathscr{M}^{p,q}} \leq \|f^\lambda\|_{\mathscr{M}^{p,q}} \leq C\lambda^\alpha\|f\|_{\mathscr{M}^{p,q}}$$

for all $f \in \mathscr{M}^{p,q}$ and all $\lambda \geq 1$, then $\alpha \geq d\mu_1(p, q)$ and $\beta \leq d\mu_2(p, q)$.

(2) There exists a constant $C > 0$ such that

$$C^{-1}\lambda^{d\mu_1(p,q)}\|f\|_{\mathscr{M}^{p,q}} \leq \|f^\lambda\|_{\mathscr{M}^{p,q}} \leq C\lambda^{d\mu_2(p,q)}\|f\|_{\mathscr{M}^{p,q}}$$

for all $f \in \mathscr{M}^{p,q}$ and all $\lambda \in (0, 1]$. Conversely, if there exist constants $C > 0$ and $\alpha, \beta \in \mathbb{R}$ such that

$$C^{-1}\lambda^\alpha\|f\|_{\mathscr{M}^{p,q}} \leq \|f^\lambda\|_{\mathscr{M}^{p,q}} \leq C\lambda^\beta\|f\|_{\mathscr{M}^{p,q}}$$

for all $f \in \mathscr{M}^{p,q}$ and all $\lambda \in (0, 1]$, then $\alpha \geq d\mu_1(p, q)$ and $\beta \leq d\mu_2(p, q)$.

We note right away that the lower bounds in Theorem 2.14 follow from the given upper bounds. For example, let $\lambda \geq 1$. Then, $1/\lambda \in (0, 1]$ and using (2) we can write

$$\|f\|_{\mathscr{M}^{p,q}} = \|(f^\lambda)^{1/\lambda}\|_{\mathscr{M}^{p,q}} \leq C\lambda^{-d\mu_2(p,q)}\|f^\lambda\|_{\mathscr{M}^{p,q}},$$

which clearly gives the left-hand side of (1). A similar argument will work for (2), by using now the upper bound in (1). Therefore, it suffices to obtain the upper bounds (and prove their optimality). The idea in [211] is to divide the proof of these upper bound estimates into the three natural cases implied by the definitions of the exponents μ_1 and μ_2. We will show in detail the argument of arguably the simplest case, while the remaining cases will only be sketched briefly.

Proof of Theorem 2.14 *(When either $(1/p, 1/q) \in \mathbb{I}_1$ or $(1/p, 1/q) \in \mathbb{I}_1^*$)* We begin by noticing that, with $\varphi = \varphi_d$, the d-dimensional Gaussian, and $\lambda > 0$ we have (see, for example, [220, Lemma 1.8]):

$$\|V_\varphi(\varphi^\lambda)\|_{L^{p,q}} = C(p, q, d)\lambda^{-d/p}(1 + \lambda^2)^{d(1/p+1q-1)/2}. \tag{2.6}$$

By Remark 1.81, immediately following Lemma 1.80, we also see that

$$\|V_{\varphi^{1/\lambda}}\varphi\|_{L^{1,1}} = \lambda^d\|V_\varphi(\varphi^\lambda)\|_{L^{1,1}}. \tag{2.7}$$

These identities, in turn, lead to the following straightforward calculation. Let $\lambda > 0$. Given $f \in \mathscr{M}^{p,q}$, and using first a change of variables, then Lemma 2.2 combined with Young's inequality for mixed Lebesgue spaces, and finally (2.6), (2.7), we get:

$$
\begin{aligned}
\|f^\lambda\|_{\mathscr{M}^{p,q}} &= \|V_\varphi(f^\lambda)\|_{L^{p,q}} = \lambda^{d(-1/p+1/q-1)}\|V_{\varphi^{1/\lambda}}f\|_{L^{p,q}} \\
&\leq \lambda^{d(-1/p+1/q-1)}\|\varphi\|_{L^2}^{-1}\|V_\varphi f\|_{L^{p,q}}\|V_{\varphi^{1/\lambda}}\varphi\|_{L^{1,1}} \\
&= \lambda^{d(-1/p+1/q-1)}\|\varphi\|_{L^2}^{-1}\|V_\varphi f\|_{L^{p,q}}\lambda^d\|V_\varphi(\varphi^\lambda)\|_{L^{1,1}} \\
&\lesssim \lambda^{d(-1/p+1/q-1)}(1+\lambda^2)^{d/2}\|f\|_{\mathscr{M}^{p,q}}.
\end{aligned}
\tag{2.8}
$$

We consider first the subcase when $(1/p, 1/q) \in \mathbb{I}_1$, and we show the upper bound estimate in (2). Note that, by definition, $\mu_2(p,q) = -1/p$. Fix $0 < \lambda \leq 1$ and $r \geq 1$. Then, by (2.8), we get

$$
\|f^\lambda\|_{\mathscr{M}^{r,1}} \lesssim \lambda^{-d/r}\|f\|_{\mathscr{M}^{r,1}}.
$$

Interpolating this estimate with the trivial one $\|f^\lambda\|_{\mathscr{M}^{2,2}} \lesssim \lambda^{-d/2}\|f\|_{\mathscr{M}^{2,2}}$, we get that

$$
\|f^\lambda\|_{\mathscr{M}^{p,q}} \lesssim \lambda^{-d\theta/2}\lambda^{-d(1-\theta)/r}\|f\|_{\mathscr{M}^{p,q}} = \lambda^{-d(\theta/2+(1-\theta)/r)}\|f\|_{\mathscr{M}^{p,q}},
$$

for all $f \in \mathscr{M}^{p,q}$, where $\theta \in [0,1]$ is such that

$$
1/p = (1-\theta)/r + \theta/2 \quad \text{and} \quad 1/q = (1-\theta) + \theta/2.
$$

Simply note now that $\theta/2 + (1-\theta)/r = 1 - 1/q + 1/p + 1/q - 1 = 1/p$ to conclude that

$$
\|f^\lambda\|_{\mathscr{M}^{p,q}} \lesssim \lambda^{-d/p}\|f\|_{\mathscr{M}^{p,q}}.
$$

We will show that the exponent $-d/p$ is in fact sharp. Assume then that, for all $f \in \mathscr{M}^{p,q}$ with $(p,q) \in \mathbb{I}_1$, and for all $\lambda \in (0,1]$, we have

$$
\|f^\lambda\|_{\mathscr{M}^{p,q}} \lesssim \lambda^\beta\|f\|_{\mathscr{M}^{p,q}}.
$$

Replacing $f = \varphi_d$, using the fact that $1/p + 1/q \geq 1$ and again (2.6), we get

$$
\begin{aligned}
\lambda^{-d/p} &\leq \lambda^{-d/p}(1+\lambda^2)^{d(1/p+1/q-1)/2} \\
&\lesssim \|\varphi^\lambda\|_{\mathscr{M}^{p,q}} \lesssim \lambda^\beta\|\varphi\|_{\mathscr{M}^{p,q}},
\end{aligned}
$$

for all $\lambda \in (0,1]$. Clearly, this implies $\beta \leq -d/p$.

Moving on now to the subcase $(1/p, 1/q) \in \mathbb{I}_1^*$, observe that $\mu_1(p, q) = -1/p$ and $(1/p', 1/q') \in \mathbb{I}_1$. Fix $\lambda \geq 1$. If $p = 1$, then $q = \infty$ and, by (2.8), we already know that

$$\|f^\lambda\|_{\mathcal{M}^{1,\infty}} \lesssim \lambda^{-2d}(1 + \lambda^2)^{d/2}\|f\|_{\mathcal{M}^{1,\infty}} \lesssim \lambda^{-d/p}\|f\|_{\mathcal{M}^{1,\infty}}.$$

If $p > 1$ and $1 < q \leq \infty$, we reduce the problem to the subcase already discussed by duality. We proceed as follows.

$$\begin{aligned}
\|f^\lambda\|_{\mathcal{M}^{p,q}} &= \sup\{|\langle f^\lambda, g\rangle| : \|g\|_{\mathcal{M}^{p',q'}} = 1\} \\
&= \lambda^{-d} \sup\{|\langle f, g^{1/\lambda}\rangle| : \|g\|_{\mathcal{M}^{p',q'}} = 1\} \\
&\lesssim \lambda^{-d} \sup\{\|f\|_{\mathcal{M}^{p,q}} \left(\frac{1}{\lambda}\right)^{-d/p'} \|g\|_{\mathcal{M}^{p',q'}} : \|g\|_{\mathcal{M}^{p',q'}} = 1\} \\
&= \lambda^{-d/p}\|f\|_{\mathcal{M}^{p,q}},
\end{aligned}$$

where in the transition to the penultimate line we used Hölder's inequality and the fact that $(1/p', 1/q') \in \mathbb{I}_1$ (which allowed us to apply the upper estimate in (2) proved in the first considered subcase for $1/\lambda \in (0, 1]$).

Finally, we show that the exponent is sharp. We revert again to a duality argument utilizing the sharpness of the exponent proved for the region $\mathbb{I}_1$. Assume then that, for all $g \in \mathcal{M}^{p,q}$ and all $\lambda \geq 1$, we have

$$\|g^\lambda\|_{\mathcal{M}^{p,q}} \lesssim \lambda^\alpha\|g\|_{\mathcal{M}^{p,q}}.$$

Suppose first that $q < \infty$, that is, $1 < p, q < \infty$. Let $f \in \mathcal{M}^{p',q'}$ and $g \in \mathcal{M}^{p,q}$ with $\|g\|_{\mathcal{M}^{p,q}} = 1$ be arbitrary. Then, using again duality as above, we get

$$\begin{aligned}
\|f^{1/\lambda}\|_{\mathcal{M}^{p',q'}} &\leq \sup_g \lambda^d \|f\|_{\mathcal{M}^{p',q'}} \|g^\lambda\|_{\mathcal{M}^{p,q}} \\
&\lesssim \sup_g \lambda^d \|f\|_{\mathcal{M}^{p',q'}} \lambda^\alpha \|g\|_{\mathcal{M}^{p,q}} = \left(\frac{1}{\lambda}\right)^{-d-\alpha}\|f\|_{\mathcal{M}^{p',q'}}.
\end{aligned}$$

Since $1/\lambda \leq 1$ and $(1/p', 1/q') \in \mathbb{I}_1$, we know already that the optimal exponent is $-d/p'$, that is, $-d - \alpha \leq -d/p'$, which is equivalent to $\alpha \geq -d/p$.

The case $q = \infty$ requires a bit more work due to the absence of a "good" duality (see the picture of the regions $\mathbb{I}_j$ and $\mathbb{I}_j^*$ with $j = 1, 2, 3$). We proceed by contradiction. Assume that for all $r \geq 1$, $\lambda \geq 1$ and $f \in \mathcal{M}^{r,\infty}$ we have

$$\|f^\lambda\|_{\mathcal{M}^{r,\infty}} \lesssim \lambda^\alpha\|f\|_{\mathcal{M}^{r,\infty}}$$

for some $\alpha < -d/r$. Interpolating again with the trivial $\mathcal{M}^{2,2}$ estimate, we obtain that, for $r < \infty$,

$$\|f^\lambda\|_{\mathscr{M}^{p,q}} \lesssim \lambda^{(\alpha r+d)(1/p-1/q)-d/p}\|f\|_{\mathscr{M}^{p,q}}$$

and, for $r = \infty$,

$$\|f^\lambda\|_{\mathscr{M}^{p,q}} \lesssim \lambda^{\alpha(1-2/q)-d/p}\|f\|_{\mathscr{M}^{p,q}};$$

here, $1/p = (1-\theta)/r + \theta/2$ and $1/q = \theta/2$ for $\theta \in (0,1)$. It is easy to see that $q \in (2,\infty)$ and $(1/p, 1/q) \in \mathbb{I}_1^*$. Since $\max\{(\alpha r+d)(1/p-1/q)-d/p, \alpha(1-2/q)-d/p\} < -d/p$, we arrive to a contradiction with the sharpness of the exponent already proved for $q < \infty$.

We will outline now the arguments that settle Theorem 2.14 in the situations where the pair $(1/p, 1/q)$ belongs to $\mathbb{I}_j \cup \mathbb{I}_j^*$, $j = 2, 3$. First, note that estimate (2.8) implies that, for all $f \in \mathscr{M}^{1,r}$, $1 \le r \le \infty$, and all $\lambda > 0$, we have

$$\|f^\lambda\|_{\mathscr{M}^{1,r}} \lesssim \lambda^{d(1/r-2)}(1+\lambda^2)^{d/2}\|f\|_{\mathscr{M}^{1,r}}.$$

Interpolating now this estimate with the trivial estimate $\|f^\lambda\|_{\mathscr{M}^{2,2}} \lesssim \lambda^{-d/2}\|f\|_{\mathscr{M}^{2,2}}$, we get

$$\|f^\lambda\|_{\mathscr{M}^{p,q}} \lesssim \lambda^{-d(2/p-1/q)}(1+\lambda^2)^{d(1/p-1/2)}\|f\|_{\mathscr{M}^{p,q}}, \tag{2.9}$$

for all $\lambda > 0$ and all $f \in \mathscr{M}^{p,q}$ with $1 \le p \le q \le \infty$ and $(1/p, 1/q) \in \mathbb{I}_2^*$.

With a little bit more effort, see [211, Lemmas 3.4–3.6], one can prove that the following estimates hold for all $f \in \mathscr{M}^{p,1}$, $1 \le p \le \infty$, and all $\lambda \ge 1$:

$$\|f^\lambda\|_{\mathscr{M}^{p,1}} \lesssim \|f\|_{\mathscr{M}^{p,1}}, \quad \text{for } p \le 2, \tag{2.10}$$

$$\|f^\lambda\|_{\mathscr{M}^{p,1}} \lesssim \lambda^{-d(2/p-1)}\|f\|_{\mathscr{M}^{p,1}}, \quad \text{for } p \ge 2. \tag{2.11}$$

Employing the usual duality trick, we then find that for all $f \in \mathscr{M}^{p,\infty}$, $1 \le p \le \infty$, and all $0 < \lambda \le 1$, we have:

$$\|f^\lambda\|_{\mathscr{M}^{p,\infty}} \lesssim \lambda^{-2d/p}\|f\|_{\mathscr{M}^{p,\infty}}, \quad \text{for } p \le 2, \tag{2.12}$$

$$\|f^\lambda\|_{\mathscr{M}^{p,\infty}} \lesssim \lambda^{-d}\|f\|_{\mathscr{M}^{p,\infty}}, \quad \text{for } p \ge 2. \tag{2.13}$$

Sketch of Proof of Theorem 2.14 (*When Either* $(1/p, 1/q) \in \mathbb{I}_2$ *or* $(1/p, 1/q) \in \mathbb{I}_2^*$) We begin with the case $(1/p, 1/q) \in \mathbb{I}_2$; here, $\mu_2(p,q) = 1/q - 1 = -1/q'$. Assume further that $1/p \le 1/q$. If $p = \infty$ and $q = 1$, the inequality we want is $\|f^\lambda\|_{\mathscr{M}^{\infty,1}} \lesssim \|f\|_{\mathscr{M}^{\infty,1}}$, $0 < \lambda \le 1$, something that we proved already (in this case, $(1/p, 1/q) \in \mathbb{I}_1$ as well). Therefore, we can assume $1 < q \le \infty$. Then, $1 \le p', q' < \infty$, $(1/p', 1/q') \in \mathbb{I}_2^*$, and $1/p' \ge 1/q'$. Let $f \in \mathscr{M}^{p,q}$ and $0 < \lambda \le 1$. Using duality and estimate (2.9), we get

$$\|f^\lambda\|_{\mathcal{M}^{p,q}} = \sup\{|\langle f^\lambda, g\rangle| : \|g\|_{\mathcal{M}^{p',q'}} = 1\}$$

$$= \lambda^{-d}\sup\{|\langle f, g^{1/\lambda}\rangle| : \|g\|_{\mathcal{M}^{p',q'}} = 1\}$$

$$\leq \lambda^{-d}\sup\{|\|f\|_{\mathcal{M}^{p,q}}\|g^{1/\lambda}\|_{\mathcal{M}^{p',q'}} : \|g\|_{\mathcal{M}^{p',q'}} = 1\}$$

$$\lesssim \lambda^{-d}\|f\|_{\mathcal{M}^{p,q}}\left(\frac{1}{\lambda}\right)^{-d(2/p'-1/q')}\left(1 + \frac{1}{\lambda^2}\right)^{d(1/p'-1/2)}$$

$$\lesssim \lambda^{-d}\lambda^{d(2/p'-1/q')}\lambda^{-2d(1/p'-1/2)}\|f\|_{\mathcal{M}^{p,q}} = \lambda^{-d/q'}\|f\|_{\mathcal{M}^{p,q}};$$

$$(2.14)$$

this is the desired estimate for $(1/p, 1/q) \in \mathbb{I}_2$, with $1/p \leq 1/q$. Note that, in particular, (2.14) gives, for $2 \leq r \leq \infty$ and $0 < \lambda \leq 1$,

$$\|f^\lambda\|_{\mathcal{M}^{r,r}} \lesssim \lambda^{-d/r'}\|f\|_{\mathcal{M}^{r,r}}.$$

By interpolating this with the estimate proved in [211, Lemma 3.5]

$$\|f^\lambda\|_{\mathcal{M}^{2,\infty}} \lesssim \lambda^{-d}\|f\|_{\mathcal{M}^{2,\infty}}, 0 < \lambda \leq 1, \tag{2.15}$$

we obtain the estimate (2.14) for $(1/p, 1/q) \in \mathbb{I}_2$ with $1/p \geq 1/q$.

Regarding the optimality of the exponent, we proceed as follows. Assume, by contradiction, that for some $\beta > -d/q'$ and for all $f \in \mathcal{M}^{p,q}$ with $(1/p, 1/q) \in \mathbb{I}_2$, $(p, q) \neq (\infty, 1)$ and all $0 < \lambda \leq 1$, we have

$$\|f^\lambda\|_{\mathcal{M}^{p,q}} \lesssim \lambda^\beta\|f\|_{\mathcal{M}^{p,q}}.$$

Fix $\epsilon \in (0, \beta + d/q')$ and consider the function

$$f(t) = \sum_{k\neq 0} |k|^{-d/q-\epsilon}e^{ik\cdot t}\varphi_d(t).$$

Then $f \in \mathcal{M}^{p,q}$ and $\|f^\lambda\|_{\mathcal{M}^{p,q}} \gtrsim \lambda^{-d/q'+\epsilon}$ for all $0 < \lambda \leq 1$; see, for example, [211, Lemma 3.8]. This implies that $-d/q' + \epsilon \geq \beta$, which is a contradiction.

The case where $(1/p, 1/q) \in \mathbb{I}_2^*$, $(p, q) \neq (1, \infty)$ can be reduced to the previously discussed case by using duality and observing $(1/p', 1/q') \in \mathbb{I}_2$.

Sketch of Proof of Theorem 2.14 (*When either $(1/p, 1/q) \in \mathbb{I}_3$ or $(1/p, 1/q) \in \mathbb{I}_3^*$*) We look first at the case $(1/p, 1/q) \in \mathbb{I}_3$. Further assume that $1/p + 1/q \geq 1$. It is easy to see that we must have $(1/p, 1/q) \in \mathbb{I}_2^*$ and $1/p \geq 1/q$. By (2.9), we get, for all $f \in \mathcal{M}^{p,q}$ and $0 < \lambda \leq 1$,

$$\|f^\lambda\|_{\mathcal{M}^{p,q}} \lesssim \lambda^{-d(2/p-1/q)}\|f\|_{\mathcal{M}^{p,q}}, \tag{2.16}$$

which is the desired estimate. We are left with the case $1/p + 1/q \leq 1$. Note that, if $q = \infty$, the estimate we are trying to prove is simply contained in (2.12). Let then $q < \infty$. Letting $p = r$ and $q = r'$, $1 \leq r \leq 2$, in (2.16), we see that for all $0 < \lambda \leq 1$,

$$\|f^{\lambda}\|_{\mathcal{M}^{r,r'}} \lesssim \lambda^{-d(3/r-1)} \|f\|_{\mathcal{M}^{r,r'}}.$$

Interpolating this with the estimate (2.15), we obtain the desired estimate (2.16) for $(1/p, 1/q) \in \mathbb{I}_3$ and $1/p + 1/q \leq 1$. For the optimality of the result, we proceed as follows. Let $1 \leq p \leq \infty$, $1 \leq q < \infty$ and $\epsilon > 0$, and let $\psi \in \mathscr{S}$ be so that $\mathrm{supp}\,(\psi) \subset [-1/2, 1/2]^d$ and $\psi \equiv 1$ on $[-1/4, 1/4]^d$. Consider the function

$$f(t) = \sum_{k \neq 0} |k|^{-d/q-\epsilon} e^{ik \cdot t} \psi(t - k).$$

Then $f \in \mathcal{M}^{p,q}$ with $\|f\|_{\mathcal{M}^{p,q}} \gtrsim \lambda^{-d(2/p-1/q)+\epsilon}$; see, for example, [211, Lemma 3.9]. Now, simply repeat the argument for the case $(1/p, 1/q) \in \mathbb{I}_2$.

Finally, the case $(1/p, 1/q) \in \mathbb{I}_3^*$, $(p, q) \neq (\infty, 1)$ follows by duality by noticing $(1/p, 1/q) \in \mathbb{I}_3$. We omit this discussion.

We end this section by pointing out that the modulation spaces behave quite well under certain nonlinear operations as well. Since this fact is of interest when studying the well posedness of nonlinear partial differential equations, we prefer to state it in the more general setting of weighted modulation spaces; see Chap. 5.

2.3 Examples

To gain a better feeling for the modulation spaces, we provide in this section some detailed calculations involving the modulation space norms of specific distributions. But first, we note that if $p \geq 1$, $q_1 \leq \min(p, p') \leq q_2$, then $\mathcal{M}^{p,q_1} \subset L^p \subset \mathcal{M}^{p,q_2}$; see [220, Proposition 1.7].

We begin with the following necessary condition for membership in certain modulation spaces.

Proposition 2.15 *Assume that* $1 \leq p \leq \infty$. *Then each* $f \in \mathcal{M}^{p,1}(\mathbb{R}^d)$ *is a continuous function.*

Proof Let $\phi \in \mathscr{S}(\mathbb{R}^d)$ be an even real-valued function with $\|\phi\|_{L^2} = 1$. Then for each $t \in \mathbb{R}^d$,

$$\left| \iint_{\mathbb{R}^{2d}} V_\phi f(x, \xi) e^{2\pi i \xi \cdot t} \phi(t - x) dx d\xi \right| \leq \|f\|_{\mathcal{M}^{p,1}} \|\phi\|_{L^{p'}}.$$

Hence, the à priori weak integral given in the inversion formula for the STFT, Proposition 2.3 (b), converges absolutely, and thus defines a function on $\mathbb{R}^d$. Consequently,

$$f(t) = \iint_{\mathbb{R}^{2d}} V_\phi f(x, \xi) e^{2\pi i \xi \cdot t} \phi(t - x) dx d\xi.$$

For each $t_0 \in \mathbb{R}^d$ and any sequence $\{t_n\}_{n \geq 1} \subset \mathbb{R}^d$ with $\lim_{n \to \infty} t_n = t_0$,

$$f(t_n) = \iint_{\mathbb{R}^{2d}} V_\phi f(x, \xi) e^{2\pi i \xi \cdot t_n} \phi(t_n - x) dx d\xi$$

$$= \int_{\mathbb{R}^d} e^{2\pi i \xi \cdot t_n} \int_{\mathbb{R}^d} V_\phi f(x, \xi) \phi(x - t_n) dx d\xi$$

$$= \int_{\mathbb{R}^d} e^{2\pi i \xi \cdot t_n} V_\phi f(\cdot, \xi) * \phi(t_n) d\xi$$

$$= \int_{\mathbb{R}^d} f_n(\xi) d\xi,$$

where $f_n(\xi) = e^{2\pi i \xi \cdot t_n} V_\phi f(\cdot, \xi) * \phi(t_n)$. Clearly,

$$\lim_{n \to \infty} f_n(\xi) = e^{2\pi i \xi \cdot t_0} V_\phi f(\cdot, \xi) * \phi(t_0)$$

pointwise, and

$$|f_n(\xi)| = |V_\phi f(\cdot, \xi) * \phi(t_n)| \leq \|V_\phi f(\cdot, \xi) * \phi\|_{L^\infty} \leq \|V_\phi f(\cdot, \xi)\|_{L^p} \|\phi\|_{L^{p'}} \in L^1.$$

We can then apply the Lebesgue dominated convergence theorem (Theorem 1.5) to conclude that $\lim_{n \to \infty} f(t_n) = f(t_0)$, which completes the proof. □

In many calculations involving modulation spaces, the d-dimensional Gaussian window $\varphi_d(x) = e^{-\pi |x|^2}$ is a preferred choice; as a signal, it belongs to all modulation spaces.

Example 2.16 Recall that, by Example 1.74, we have $|V_{\varphi_d} \varphi_d(x, \xi)| = 2^{-d/2} \varphi_d(x) \varphi_d(\xi)$. This immediately gives that $\varphi_d \in \mathcal{M}^{p,q}$ for all $0 < p, q \leq \infty$.

Example 2.17 Let $a \in \mathbb{R}^d$. By Example 1.78, we have that $|V_{\varphi_d} \delta_0(x, \xi)| = \varphi_d(x)$. Further combining this with Example 1.34, we see that $\delta_a \in \mathcal{M}^{p,\infty}$ for $0 < p \leq \infty$. However, $\delta_a \notin \mathcal{M}^{p,q}$ for any $q > 0, q \neq \infty$.

Example 2.18 By Example 1.79, we know that $|V_{\varphi_d} 1(x, \xi)| = \varphi_d(\xi)$. This gives $1 \in \mathcal{M}^{\infty,q}$ for $0 < q \leq \infty$ and $1 \notin \mathcal{M}^{p,q}$ for any $p > 0, p \neq \infty$.

Example 2.19 Let $f(x) = e^{i\pi x^2}$, $x \in \mathbb{R}$. By Example 1.75, $|V_{\varphi_1} f(x, \xi)| \sim e^{-\pi(x-\xi)^2/2}$. Thus, for all $p, q > 0$ with $q \neq \infty$, $f \in \mathcal{M}^{p,\infty} \setminus \mathcal{M}^{p,q}$.

Example 2.20 In Example 1.76 we defined $\sigma_{t;2}(x) = e^{i\pi t|x|^2}$, $x \in \mathbb{R}^d$ and $t \in \mathbb{R}$. Furthermore, we showed there that $|V_{\varphi_d}\sigma_{t;2}(x,\xi)| = (1+t^2)^{-d/4}e^{-\frac{\pi}{1+t^2}|tx-\xi|^2}$. Let $p, q > 0$, $p \neq \infty$. It is now easy to compute

$$\|\sigma_{t;2}\|_{\mathscr{M}^{p,\infty}} = \sup_{\xi}\left(\int_{\mathbb{R}^d}|V_{\varphi_d}\sigma_{t;2}(x)|^p\,dx\right)^{1/p} = (1+t^2)^{-d/4}\left(\int_{\mathbb{R}^d}e^{-\frac{\pi pt^2|x|^2}{1+t^2}}\,dx\right)^{1/p}.$$

Since $\int_{\mathbb{R}^d}e^{-\pi a|x|^2}\,dx = a^{-d/2}$, we obtain $\|\sigma_{t;2}\|_{\mathscr{M}^{p,\infty}} = p^{-d/2p}(1+t^2)^{d(2-p)/4p}t^{-d/p}$. The previous calculation gives that $\sigma_{t,2} \in \mathscr{M}^{p,\infty} \setminus \mathscr{M}^{p,q}$ for $q \neq \infty$. A similar computation yields $\|\sigma_{t;2}\|_{W(\mathscr{F}L^p, L^\infty)} = p^{-d/2p}(1+t^2)^{d(2-p)/4p}$.

Example 2.21 Let $1 \leq p, q \leq \infty$. Let $\mathbb{B} = \overline{\mathbb{B}(0,1)}$ be the closed unit ball centered at the origin in $\mathbb{R}^d$, and $\chi_{\mathbb{B}}$ its characteristic function. From Proposition 2.9 it follows that $\chi_{\mathbb{B}} \in (\mathscr{F}L^q)_{\text{comp}}(\mathbb{R}^d)$. But

$$\widehat{\chi_{\mathbb{B}}}(\xi) = \frac{J_{\frac{d}{2}}(2\pi|\xi|)}{|\xi|^{d/2}},$$

where $J_{\frac{d}{2}}$ is the Bessel function of order $d/2$; see [119, 120]. Following [119, Appendix B-6, B-7], it is known that for $0 < |\xi| < 1$, $|J_{\frac{d}{2}}(2\pi|\xi|)| \lesssim |\xi|^{d/2}$, and for $|\xi| > 1$, $|J_{\frac{d}{2}}(2\pi|\xi|)| \lesssim |\xi|^{-1/2}$. Using these asymptotics, we conclude that $\widehat{\chi_{\mathbb{B}}} \in L^q(\mathbb{R}^d)$ for $q > 2 - \frac{2}{d+1}$ for all $d > 1$. Consequently, for all $d \geq 1$, $\chi_{\mathbb{B}} \in \mathscr{M}^{1,1+\epsilon}_{\text{comp}}(\mathbb{R}^d)$ for all $\epsilon > \frac{d-1}{d+1}$. In particular, for $d = 1$, $\chi_{\mathbb{B}} \in \mathscr{M}^{1,1+\epsilon}_{\text{comp}}(\mathbb{R})$ for all $\epsilon > 0$. Note that since $\chi_{\mathbb{B}}$ is discontinuous, then for every $1 \leq p \leq \infty$, $\chi_{\mathbb{B}} \notin \mathscr{M}^{p,1}(\mathbb{R}^d)$. Thus, the result $\chi_{\mathbb{B}} \in \mathscr{M}^{1,1+\epsilon}_{\text{comp}}(\mathbb{R})$ is optimal in dimension $d = 1$.

Example 2.22 For $d = 1$, let $f = \chi_{[0,1]} * \chi_{[0,1]}$. By Example 1.49, we know that

$$\widehat{f}(\xi) = \left(\frac{1 - e^{-2\pi i\xi}}{2\pi i\xi}\right)^2.$$

f is compactly supported and it is easy to see that $\widehat{f} \in L^1$, hence $f \in (\mathscr{F}L^1)_{\text{comp}}(\mathbb{R})$, which implies that $f \in \mathscr{M}^{1,1}(\mathbb{R})$.

Example 2.23 Let $f(x) = (1 - |x|)^{1/2}$ for $|x| < 1$ and $f(x) = 0$ otherwise. A direct calculation shows that $f^2 \in \mathscr{M}^{1,1}$. We claim that $f \notin \mathscr{M}^{1,1}$. To prove the claim, we interpret f as a function in $L^2(-1, 1)$ and show that its Fourier series does not converge at 1; in particular, this shows that f cannot be in $\mathscr{F}L^1(\mathbb{R})$, and hence $f \notin \mathscr{M}^{1,1}$. Going back to the Fourier series of f restricted to $(-1, 1)$, let us denote by S_N its partial sum of order N. We have $S_N = f * D_N$, where D_N is the Dirichlet kernel. A repeated integration by parts shows that $f * D_N(1) \not\to 0$ as $N \to \infty$, thus proving the claim; see [119].

2.4 The Feichtinger Algebra

In this section we focus on the modulation space $\mathscr{M}^{1,1}(\mathbb{R}^d)$ which is also known as *the Feichtinger algebra* and sometimes denoted S_0. The Feichtinger algebra and its dual $\mathscr{M}^{\infty,\infty}(\mathbb{R}^d)$ represent natural classes of Banach spaces which are amenable to a variety of harmonic analysis questions. Additionally, they have the advantage of being equipped with norms (as opposed to the pair of dual spaces $\mathscr{S}(\mathbb{R}^d)$ and $\mathscr{S}'(\mathbb{R}^d)$ which stem from families of seminorms); see, for example, [98] and [75] for a discussion of Banach Gelfand triples in analysis. Moreover, we will prove that windows in $\mathscr{M}^{1,1}$ are natural choices to define the modulation space norms and we will introduce a powerful discretization of the modulation spaces that relies on some essential properties of $\mathscr{M}^{1,1}$. For more on the Feichtinger algebra and its properties we refer to [26, 157].

Proposition 2.24 *The modulation space $\mathscr{M}^{1,1}$ satisfies the following properties:*

(a) *It is invariant under the Fourier transform.*

(b) *It is an algebra under pointwise multiplication.*

(c) *It is an algebra under convolution.*

(d) $(\mathscr{F}L^1)_{comp}(\mathbb{R}^d) = \mathscr{M}^{1,1}_{comp}(\mathbb{R}^d) \subset \mathscr{M}^{1,1}(\mathbb{R}^d)$.

(e) *Let $1 \leq p \leq \infty$. Given $g \in \mathscr{M}^{1,1}(\mathbb{R}^d)$, the linear map $M_g : \mathscr{M}^{p,p}(\mathbb{R}^d) \to \mathscr{M}^{p,p}(\mathbb{R}^d)$ given by $M_g f = fg$ is bounded. Similarly, the linear map $C_g : \mathscr{M}^{p,p}(\mathbb{R}^d) \to \mathscr{M}^{p,p}(\mathbb{R}^d)$ given by $C_g f = f * g$ is bounded.*

(f) $f \in \mathscr{M}^{1,1}(\mathbb{R}^d)$ *if and only if for some $\phi \in \mathscr{M}^{1,1}(\mathbb{R}^d)$ (and hence for all $\phi \in \mathscr{M}^{1,1}(\mathbb{R}^d)$), $V_\phi f \in L^1(\mathbb{R}^{2d})$. In particular, $f \in \mathscr{M}^{1,1}$ if and only if $V_f f \in L^1(\mathbb{R}^{2d})$.*

Proof

(a) This was already proved in Proposition 2.12 (a) (for $p = q = 1$). In addition, observe that if $f \in \mathscr{M}^{1,1}(\mathbb{R}^d)$, then $\hat{f} \in \mathscr{M}^{1,1}(\mathbb{R}^d)$ and

$$\|\hat{f}\|_{\mathscr{M}^{1,1}(\mathbb{R}^d)} \sim \|f\|_{\mathscr{M}^{1,1}(\mathbb{R}^d)}.$$

(b) Let $f, g \in \mathscr{M}^{1,1}(\mathbb{R}^d)$. We need to prove that $fg \in \mathscr{M}^{1,1}$ and $\|fg\|_{\mathscr{M}^{1,1}} \leq \|f\|_{\mathscr{M}^{1,1}} \|g\|_{\mathscr{M}^{1,1}}$. For this, let $\phi_k \in \mathscr{S}(\mathbb{R}^d)$, $k = 1, 2$ and define $\phi = \phi_1 \phi_2$. A series of straightforward calculations gives

$$V_\phi(fg)(x, \xi) = \widehat{f \cdot T_x \overline{\phi_1}} * \widehat{g \cdot T_x \overline{\phi_2}}(\xi), \qquad (2.17)$$

which yields

$$\|V_\phi(fg)\|_{L^{1,1}(\mathbb{R}^{2d})} = \iint_{\mathbb{R}^{2d}} |V_\phi(fg)(x, \xi)| \, dx \, d\xi$$

$$= \iint_{\mathbb{R}^{2d}} |\widehat{f \cdot T_x \overline{\phi_1}} * \widehat{g \cdot T_x \overline{\phi_2}}(\xi)| \, d\xi \, dx$$

$$\leq \int_{\mathbb{R}^d} \|\widehat{f \cdot T_x\overline{\phi_1}}\|_{L^1} \|\widehat{g \cdot T_x\overline{\phi_2}}\|_{L^1} dx$$

$$\leq \sup_x \|\widehat{f \cdot T_x\overline{\phi_1}}\|_{L^1} \int_{\mathbb{R}^d} \|\widehat{g \cdot T_x\overline{\phi_2}}\|_{L^1} dx$$

$$= \sup_x \|\widehat{f \cdot T_x\overline{\phi_1}}\|_{L^1} \|g\|_{\mathscr{M}^{1,1}}$$

$$\leq \|f\|_{\mathscr{M}^{\infty,1}} \|g\|_{\mathscr{M}^{1,1}}$$

$$\leq \|f\|_{\mathscr{M}^{1,1}} \|g\|_{\mathscr{M}^{1,1}},$$

where we have used the fact that $\mathscr{M}^{1,1} \subset \mathscr{M}^{\infty,1}$.

(c) For $f, g \in \mathscr{M}^{1,1}$, since $\widehat{f * g} = \widehat{f}\widehat{g}$, and by parts (a), (b), we obtain that $\widehat{f}\widehat{g} \in \mathscr{M}^{1,1}$. Consequently, $\widehat{f * g} \in \mathscr{M}^{1,1}$, thus $f * g \in \mathscr{M}^{1,1}$. The norm estimates easily follow.

(d) This was proved in Proposition 2.9.

(e) Let $f \in \mathscr{M}^{1,1}$, $g \in \mathscr{M}^{p,q}$ and $\phi = \phi_1\phi_2$ as in part (b). Using (2.17) we have

$$\|V_\phi(fg)\|_{L^{p,p}(\mathbb{R}^{2d})} = \left(\int_{\mathbb{R}^d} \int_{\mathbb{R}^d} |\widehat{f \cdot T_x\overline{\phi_1}} * \widehat{g \cdot T_x\overline{\phi_2}}(\xi)|^p dxd\xi \right)^{1/p}$$

$$\leq \left(\int_{\mathbb{R}^d} \|\widehat{f \cdot T_x\overline{\phi_1}}\|_{L^1}^p \|\widehat{g \cdot T_x\overline{\phi_2}}\|_{L^p}^p dx \right)^{1/p}$$

$$\leq \sup_{x \in \mathbb{R}^d} \|\widehat{f \cdot T_x\overline{\phi_1}}\|_{L^1} \|V_{\phi_2} g\|_{L^{p,p}}$$

$$\leq \|f\|_{\mathscr{M}^{1,1}} \|g\|_{\mathscr{M}^{p,p}}.$$

The last part follows from Proposition 2.12 and the boundedness of M_g.

(f) Fix $\phi_0 \in \mathscr{S}$ and let $f, \phi \in \mathscr{M}^{1,1}$. By Lemma 2.2 with $C = |\langle \phi_0, \phi \rangle|^{-1}$, we have

$$|V_\phi f(x, \xi)| \leq C|V_\phi\phi_0| * |V_{\phi_0} f|(x, \xi). \tag{2.18}$$

The result then follows from Young's inequality.

$\square$

The next result proves that within the family of modulation spaces, $\mathscr{M}^{1,1}$ is rich enough, in the sense that it is dense in all other modulation spaces.

Proposition 2.25 *If* $1 \leq p, q < \infty$, *then* $\mathscr{M}^{1,1}(\mathbb{R}^d)$ *is dense in* $\mathscr{M}^{p,q}(\mathbb{R}^d)$.

Proof First, observe that $\mathscr{M}^{1,1} \subset \mathscr{M}^{p,q}$ as can be seen by applying Young's inequality to (2.18).

Next, let $1 \leq p, q < \infty$, $f \in \mathcal{M}^{p,q}(\mathbb{R}^d)$, $\phi \in \mathscr{S}(\mathbb{R}^d)$ such that $\|\phi\|_{L^2} = 1$. For each $n \geq 1$ we define $F_n = V_\phi f \cdot \chi_{\mathbb{K}_n}$, and $f_n = V_\phi^* F_n$ as in the proof of Theorem 2.5. It is clear that $f_n \in \mathcal{M}^{1,1}(\mathbb{R}^d)$ and that $\lim_{n\to\infty} \|V_\phi f - F_n\|_{L^{p,q}} = 0$, and the rest of the proof is similar to that of Theorem 2.5. $\square$

Recall that we proved in Proposition 2.3 that the definition of the modulation spaces is independent of the choice of the window function $\phi \in \mathscr{S}(\mathbb{R}^d)$. In fact, we can prove that this remains true even when the window is chosen in $\mathcal{M}^{1,1}(\mathbb{R}^d)$.

Proposition 2.26 *Let* $1 \leq p, q \leq \infty$, $\phi_1, \phi_2 \in \mathcal{M}^{1,1}(\mathbb{R}^d)$ *be given. There exist constants* $C_1, C_2 > 0$ *such that for each* $f \in \mathcal{M}^{p,q}(\mathbb{R}^d)$,

$$C_1 \|V_{\phi_1} f\|_{L^{p,q}} \leq \|V_{\phi_2} f\|_{L^{p,q}} \leq C_2 \|V_{\phi_1} f\|_{L^{p,q}}.$$

Consequently,

$$\|f\|_{\mathcal{M}^{p,q}} \sim \|V_{\phi_1} f\|_{L^{p,q}}.$$

Proof The argument is similar to the one proving Proposition 2.3. $\square$

2.5 Gabor Frames and Feichtinger's Algebra

We now give a discrete characterization of the modulation spaces using Gabor frames. These systems were originally introduced by D. Gabor in 1946 [111]. Subsequently, the subject of Gabor analysis in $L^2(\mathbb{R}^d)$ has attracted a lot of attention especially during the 1990s where they were investigated along with wavelets. There are extensive treatises on Gabor frames; see, for example, [53, 87–89, 101, 103–105, 125, 130, 145, 148, 199, 230]. For a brief introduction to Gabor analysis and some interesting open problems in the area, including the fascinating HRT conjecture of Heil-Ramanathan-Topiwala [146], we refer the reader to [195]. Here, we only state the decomposition of functions in the modulation spaces, including Feichtinger's algebra, via the summability of their Gabor coefficients.

Theorem 2.27 *Let* $\alpha, \beta > 0$ *be such that* $\alpha\beta < 1$. *Suppose that* $\phi \in \mathcal{M}^{1,1}(\mathbb{R}^d)$ *is such that* $\{M_{n\beta} T_{m\alpha} \phi\}_{n,m \in \mathbb{Z}^d}$ *is a Gabor frame for* $L^2(\mathbb{R}^d)$, *that is, there exist,* $0 < A \leq B < \infty$ *such that for all* $f \in L^2(\mathbb{R}^d)$,

$$A\|f\|_2^2 \leq \sum_{n,m \in \mathbb{Z}^d} |\langle f, M_{n\beta} T_{m\alpha} \phi \rangle|^2 \leq B\|f\|_2^2. \tag{2.19}$$

Then $f \in \mathcal{M}^{1,1}(\mathbb{R}^d)$ *if and only if*

$$C_\phi f = \{\langle f, M_{n\beta} T_{m\alpha} \phi \rangle\}_{(m,n) \in \mathbb{Z}^d \times \mathbb{Z}^d} \in \ell^{1,1}(\mathbb{Z}^{2d})$$

and

$$\|f\|_{\mathscr{M}^{1,1}} \sim \|C_\phi f\|_{\ell^{1,1}}. \tag{2.20}$$

Moreover, there exists $\gamma \in \mathscr{M}^{1,1}(\mathbb{R}^d)$ *such that*

$$f = \sum_{m\in\mathbb{Z}^d} \sum_{n\in\mathbb{Z}^d} \langle f, M_{n\beta} T_{m\alpha}\phi\rangle M_{n\beta} T_{m\alpha}\gamma = \sum_{m\in\mathbb{Z}^d} \sum_{n\in\mathbb{Z}^d} \langle f, M_{n\beta} T_{m\alpha}\gamma\rangle M_{n\beta} T_{m\alpha}\phi,$$
$$\tag{2.21}$$

with the series convergent in the norm of $\mathscr{M}^{1,1}$.

Consequently, for $1 \le p,q \le \infty$, $f \in \mathscr{M}^{p,q}(\mathbb{R}^d)$ *if and only if*

$$C_\phi f = \{\langle f, M_{n\beta} T_{m\alpha}\phi\rangle\}_{(m,n)\in\mathbb{Z}^d\times\mathbb{Z}^d} \in \ell^{p,q}(\mathbb{Z}^{2d})$$

and

$$\|f\|_{\mathscr{M}^{p,q}} \sim \|C_\phi f\|_{\ell^{p,q}}. \tag{2.22}$$

Moreover,

$$f = \sum_{m\in\mathbb{Z}^d} \sum_{n\in\mathbb{Z}^d} \langle f, M_{n\beta} T_{m\alpha}\phi\rangle M_{n\beta} T_{m\alpha}\gamma = \sum_{m\in\mathbb{Z}^d} \sum_{n\in\mathbb{Z}^d} \langle f, M_{n\beta} T_{m\alpha}\gamma\rangle M_{n\beta} T_{m\alpha}\phi,$$
$$\tag{2.23}$$

with the series convergent in the norm of $\mathscr{M}^{p,q}$, *when* $p,q < \infty$. *In the case* $p = \infty$ *or* $q = \infty$ *the convergence is in the weak sense.*

Remark 2.28

(a) This result was first established in [101] under the assumption that $\alpha\beta \in \mathbb{Q}$, while the irrational case ($\alpha\beta \notin \mathbb{Q}$) was proved later in [130].
(b) We observe that in order to guarantee a Gabor expansion for all functions in the modulation spaces $\mathscr{M}^{p,q}$ one needs only to assume that the window ϕ belongs to the Feichtinger algebra $\mathscr{M}^{1,1}$ and it generates a Gabor frame for $L^2(\mathbb{R}^d)$. In a way, this amounts to saying that the invertibility of the Gabor frame operator

$$Sf := S_\phi^{\alpha,\beta} f = \sum_{k,n\in\mathbb{Z}^d} \langle f, M_{n\beta} T_{m\alpha}\phi\rangle M_{n\beta} T_{m\alpha}\phi$$

on $L^2(\mathbb{R}^d)$ along with the choice of the window in the Feichtinger algebra is sufficient for its invertibility on all modulation spaces. Extensions of this result to other spaces can be found in [1, 128, 132, 173].

We conclude this section with a result needed in our discussion about the boundedness of multipliers in Chap. 4.

Proposition 2.29 *Let $\phi \in \mathcal{M}^{1,1}$ and $1 \leq p, q \leq \infty$. Then there exists a constant $C > 0$ such that for all $\mathbf{a} = (a_n)_{n \in \mathbb{Z}^d} \in \ell^p(\mathbb{Z}^d)$ and $\mathbf{b} = (b_n)_{n \in \mathbb{Z}^d} \in \ell^q(\mathbb{Z}^d)$ we have*

$$\Big\| \sum_{n \in \mathbb{Z}^d} a_n T_n \phi \Big\|_{\mathcal{M}^{p,q}} \leq C \|\mathbf{a}\|_{\ell^p} \quad and \quad \Big\| \sum_{n \in \mathbb{Z}^d} b_n T_n \phi \Big\|_{\mathcal{M}^{p,q}} \leq C \|\mathbf{b}\|_{\ell^q}.$$

If $p < \infty$ and $q < \infty$, then both sums converge unconditionally in $\mathcal{M}^{p,q}$. If $p = \infty$ or $q = \infty$, and $(p, q) \notin \{(1, \infty), (\infty, 1)\}$, then both sums converge weak* in $\mathcal{M}^{p,q}$. If $p = \infty$ or $q = \infty$, and $(p, q) \in \{(1, \infty), (\infty, 1)\}$, then both sums converge weak* in $\mathcal{M}^{\infty,\infty}$.

Proof One simply needs to adapt the arguments of [125, Theorem 12.2.4] to this particular case. □

2.6 Notes

The modulation space $\mathcal{M}^{1,1}(\mathbb{R}^d)$ was first introduced by Feichtinger in [93] under the name of Segal algebra and was denoted S_0. Later, Feichtinger introduced a family of function spaces called Wiener-type spaces that are now known as the modulation spaces. For a historical perspective on this topic we refer the interested reader to [94] and [96], and to [97] for a discussion about the place of modulation spaces in the scale of functions spaces used in analysis. For more on the Feichtinger algebra we refer to [157].

It is not an understatement to say that modulation spaces are here to stay, and for good reasons. We wish to end this brief section with two quotes which carry weight with those interested in this area of time-frequency analysis. Moreover, in the following sections of this manuscript, we will try to follow closely the principles outlined in them.

> From the very beginning modulation spaces have been treated in analogy to the corresponding theory of Besov spaces, essentially by replacing the dyadic decompositions of unity by uniform ones. One of the reasons to do this was the construction of smoothness spaces over general LCA groups which do not have dilation. This is of course possible by replacing dilation with frequency shifts. (Feichtinger [94])

> Modulation spaces have perhaps not received as much attention in time-frequency analysis as they deserve. Only recently has it been fully recognized that the modulation spaces are the right function spaces for time-frequency analysis and that they occur in many problems in the same way as Besov spaces are attached to wavelet theory. (Gröchenig [125])

Chapter 3
Equivalent Definitions of Modulation Spaces

In this chapter, we give Feichtinger's original definition of the modulation spaces $\mathcal{M}^{p,q}$ when the indices p, q are restricted to the range $[1, \infty]$. This restriction is of course very convenient from many points of view; recall, for example, that with the appropriately defined norms (3.1) they are Banach spaces. However, in applications to partial differential equations or pseudodifferential calculus, it has become increasingly important to extend the definition of modulation spaces to indices p, q that are positive but below 1. There is nothing stopping us to define $\mathcal{M}^{p,q}$ for the range of indices $p, q \in (0, \infty]$. We are now dealing with a complete quasi-norm defined by (3.1). Galperin and Samarah [114] extended Feichtinger's original definition of modulation spaces to the full range of indices $p, q \in (0, \infty]$ and, later on, Kobayashi [166, 167] rediscovered these spaces as well. In the following, we use his approach to give an equivalent definition of the modulation spaces and show that it agrees with Feichtinger's when one restricts the indices p, q to $[1, \infty]$. We note that this is different from the co-orbit theory that Rauhut [198] used to define the modulation spaces for the full range $0 < p, q \le \infty$. Finally, we observe that the cases where one or both of the indices are equal to 0 were already investigated and applied to obtain the compactness of some special types of operators in [18, Definition 2.1, Lemma 2.2, Proposition 2.3] and [106, Lemma 3.8, Theorem 3.15].

3.1 The Original Definition of the Modulation Spaces: Similarity with Besov Spaces

We will explore here other (equivalent) definitions of modulation spaces and their basic properties that are amenable to the range of indices $0 < p, q \le \infty$. To distinguish them from the definition of $\mathcal{M}^{p,q}$ spaces given in Definition 2.1, we will denote the, a priori distinct, spaces obtained via the norms in [166] by $M^{p,q}$.

© Springer Science+Business Media, LLC, part of Springer Nature 2020
Á. Bényi, K. A. Okoudjou, *Modulation Spaces*, Applied and Numerical
Harmonic Analysis, https://doi.org/10.1007/978-1-0716-0332-1_3

A posteriori, and of importance to us, the new definitions recover the classical one, that is, $M^{p,q} = \mathscr{M}^{p,q}$ for all $0 < p, q \leq \infty$. As we will indicate below, the formulations we discuss here are rooted in Triebel's idea of developing the theory of modulation spaces in the same spirit as that of the Besov spaces, see [228]. For further details on the parallels between the modulation spaces and other function spaces, we refer to [225, 226].

Definition 3.1 Let $0 < p, q \leq \infty$, and $\phi \in \mathscr{S}(\mathbb{R}^d)$ be (fixed) such that supp $(\widehat{\phi}) \subset \overline{\mathbb{B}(0, 1)}$ and (for sufficiently small $\alpha > 0$) $\sum_{k\in\mathbb{Z}^d} \widehat{\phi}(\xi - \alpha k) = 1$ for all $\xi \in \mathbb{R}^d$. The modulation space $M^{p,q} = M^{p,q}(\mathbb{R}^d)$ is the set of all tempered distribution $f \in \mathscr{S}'(\mathbb{R}^d)$ for which

$$\|f\|_{M^{p,q}} := \left(\int_{\mathbb{R}^d} \left(\int_{\mathbb{R}^d} |(f * M_\xi \phi)(x)|^p dx \right)^{q/p} d\xi \right)^{1/q} < \infty. \tag{3.1}$$

When $p = \infty$ or $q = \infty$, the essential supremum is used.

It is useful to think of the expression above as the following identity

$$\|f\|_{M^{p,q}} = \left(\int_{\mathbb{R}^d} \|f * M_\xi \phi\|_{L^p}^q d\xi \right)^{1/q} = \|(f * M_\xi \phi)(x)\|_{L_x^p L_\xi^q}. \tag{3.2}$$

In general, $\| \cdot \|_{M^{p,q}}$ in (3.2) defines only a quasi-norm.

Recall now that Feichtinger's definition of modulation spaces is independent of the choice of a window in $\mathscr{S}(\mathbb{R}^d) \setminus \{0\}$. When $p, q \in (0, \infty]$, $\mathscr{M}^{p,q}$ is also only a quasi-Banach under the quasi-norm defined by (3.1),

$$\|f\|_{\mathscr{M}^{p,q}} = \|V_\phi f(x, \xi)\|_{L_x^p L_\xi^q}. \tag{3.3}$$

Our first observation is that, under the appropriate assumption on the window function, the quasi-norms in (3.2) and (3.3) are the same.

Proposition 3.2 *Let $0 < p, q \leq \infty$ and let $\phi \in \mathscr{S}$ be as in Definition 3.1. For all $f \in \mathscr{S}'$ we have $\|f\|_{M^{p,q}} = \|f\|_{\mathscr{M}^{p,q}}$. In particular, $M^{p,q} = \mathscr{M}^{p,q}$ for $1 \leq p, q \leq \infty$.*

Proof The statement is an easy consequence of the last stated equivalent form of the STFT in (1.12), $V_\phi f(x, \xi) = e^{-2\pi i x \cdot \xi}(f * M_\xi \overleftarrow{\widetilde{\phi}})(x)$. □

Returning now to the new Definition 3.1, we see that the window function ϕ is assumed to satisfy some additional conditions; we will write $\phi \in \mathscr{S}_\alpha(\mathbb{R}^d)$ for such a window. More specifically, we have the following definition; the assumption "sufficiently small" is necessary to guarantee the nonemptiness of this space of windows.

Definition 3.3 Fix $\alpha > 0$ to be sufficiently small. We say that $\phi \in \mathscr{S}_\alpha = \mathscr{S}_\alpha(\mathbb{R}^d)$ if $\phi \in \mathscr{S}(\mathbb{R}^d)$, supp $(\widehat{\phi}) \subset \overline{\mathbb{B}(0, 1)}$ and $\sum_{k\in\mathbb{Z}^d} \widehat{\phi}(\xi - \alpha k) = 1$ for all $\xi \in \mathbb{R}^d$.

The next result [166, Theorem 3.3] provides an equivalent quasi-norm for the modulation space $M^{p,q}$.

Theorem 3.4 *Let $0 < p, q \leq \infty$, and $\phi \in \mathscr{S}_\alpha$, with $\alpha > 0$ sufficiently small, be fixed. Then*

$$|||f|||_{M^{p,q}} := \left(\sum_{k \in \mathbb{Z}^d} \left(\int_{\mathbb{R}^d} |(f * M_{\alpha k}\phi)(x)|^p dx \right)^{q/p} \right)^{1/q} \tag{3.4}$$

is an equivalent quasi-norm on $M^{p,q}$. When $p = \infty$ or $q = \infty$, the essential supremum is used.

Note that, in view of the discussion in Sect. 1.1.3, the quasi-norm in (3.4) can be rewritten as

$$|||f|||_{M^{p,q}} = \|(f * M_{\alpha k}\phi)_{k \in \mathbb{Z}^d}\|_{L^p \ell^q}. \tag{3.5}$$

One should contrast the previous definition of modulation spaces via the quasi-norm (3.5) with the one of Besov spaces; see Chap. 6.

The main idea behind the new approach to modulation spaces detailed here is to replace the dilation operation in the expression of the $B_0^{p,q}$-quasi-norm given by (6.1) with that of modulation in (3.5).

Proof We need to show that $\|f\|_{M^{p,q}} \lesssim |||f|||_{M^{p,q}} \lesssim \|f\|_{M^{p,q}}$.

The first estimate Fix $\xi \in \mathbb{R}^d$ and let $k = (k_j)_{j=1}^d \in \mathbb{Z}^d$ be such that $\xi \in [\alpha k, \alpha(k+1)]$, where we denoted $[\alpha k, \alpha(k+1)] = \prod_{j=1}^d [\alpha k_j, \alpha(k_j+1)]$. Since $\phi \in \mathscr{S}_\alpha$, we can find an $N = N(\alpha, d) \in \mathbb{N}$ such that

$$f * M_\xi \phi = f * \mathscr{F}^{-1}(T_\xi \widehat{\phi})$$

$$= f * \left(T_\xi \widehat{\phi} \sum_{|l| \leq N} T_{\alpha(k+l)} \widehat{\phi} \right)^\vee$$

$$= f * M_\xi \phi * \sum_{|l| \leq N} M_{\alpha(k+l)}\phi. \tag{3.6}$$

In the equalities above, we used repeatedly some of the basic properties of the Fourier transform, see Proposition 1.46. Next, we prove that, for all $p > 0$, we have the estimate

$$\|f * M_{\alpha(k+l)}\phi * M_\xi \phi\|_{L^p} \lesssim \|f * M_{\alpha(k+l)}\phi\|_{L^p}, \tag{3.7}$$

where the implicit constant in the right-hand side is $C = \|\phi\|_{L^{\max(p,1)}}$. We distinguish two cases. If $p \geq 1$, we use Young's inequality (1.7) and the fact that $M_\xi \phi \in L^q$ for all $q > 0$ to get

$$\|f * M_{\alpha(k+l)}\phi * M_\xi\phi\|_{L^p} \leq \|f * M_{\alpha(k+l)}\phi\|_{L^p}\|M_\xi\phi\|_{L^1} \lesssim \|f * M_{\alpha(k+l)}\phi\|_{L^p}.$$

If $p \in (0, 1)$, we apply the obvious point-wise inequality $|f * g| \leq |f| * |g|$ and Proposition 1.58 to obtain

$$\|f * M_{\alpha(k+l)}\phi * M_\xi\phi\|_{L^p} \leq \||f * M_{\alpha(k+l)}\phi| * |M_\xi\phi|\|_{L^p} \leq \|f * M_{\alpha(k+l)}\phi\|_{L^p}\|\phi\|_{L^p}$$

$$\lesssim \|f * M_{\alpha(k+l)}\phi\|_{L^p}.$$

A combination of (3.6), the triangle inequality, Lemma 1.14, and (3.7) (in this order) gives

$$\|f * M_\xi\phi\|_{L^p}^p \doteq \int_{\mathbb{R}^d} \left| \sum_{|l| \leq N} f * M_{\alpha(k+l)}\phi * M_\xi\phi(x) \right|^p dx$$

$$\leq \int_{\mathbb{R}^d} \left(\sum_{|l| \leq N} |f * M_{\alpha(k+l)}\phi * M_\xi\phi(x)| \right)^p dx$$

$$\lesssim \int_{\mathbb{R}^d} \sum_{|l| \leq N} |f * M_{\alpha(k+l)}\phi * M_\xi\phi(x)|^p dx$$

$$\lesssim \sum_{|l| \leq N} \|f * M_{\alpha(k+l)}\phi * M_\xi\phi\|_{L^p}^p$$

$$\lesssim \sum_{|l| \leq N} \|f * M_{\alpha(k+l)}\phi\|_{L^p}^p. \tag{3.8}$$

By raising the previous estimate (3.8) to the q/p power, integrating over $[\alpha k, \alpha(k + 1)]$, summing over $k \in \mathbb{Z}^d$ and using again Lemma 1.14 we obtain the following sequence of estimates:

$$\int_{\mathbb{R}^d} \|f * M_\xi\phi\|_{L^p}^q d\xi \lesssim \sum_{k \in \mathbb{Z}^d} \int_{[\alpha k, \alpha(k+1)]} \left(\sum_{|l| \leq N} \|f * M_{\alpha(k+l)}\phi\|_{L^p}^p \right)^{q/p} d\xi$$

$$\lesssim \sum_{k \in \mathbb{Z}^d} \int_{[\alpha k, \alpha(k+1)]} \sum_{|l| \leq N} (\|f * M_{\alpha(k+l)}\phi\|_{L^p}^p)^{q/p} d\xi$$

$$\lesssim \sum_{k \in \mathbb{Z}^d} \|f * M_{\alpha k}\phi\|_{L^p}^q. \tag{3.9}$$

Finally, by raising the estimate (3.9) to the $1/q$ power gives the first estimate, $\|f\|_{M^{p,q}} \lesssim \||f\||_{M^{p,q}}$.

The second estimate Fix $k \in \mathbb{Z}^d$ and let $\xi \in [\alpha k, \alpha(k+1)]$ be arbitrary. We proceed similarly as in the calculation leading to (3.6) and obtain

$$f * M_{\alpha k}\phi = \sum_{|l|\leq N} f * M_{\xi+\alpha l}\phi * M_{\alpha k}\phi.$$

Also, as for (3.7), we can prove that we have the following estimate:

$$\|f * M_{\xi+\alpha l}\phi * M_{\alpha k}\phi\|_{L^p} \lesssim \|f * M_{\xi+\alpha l}\phi\|_{L^p}.$$

An argument mimicking that of (3.8), which uses the last inequality and the identity preceding it, the triangle inequality and Lemma 1.14, gives now

$$\|f * M_{\alpha k}\phi\|_{L^p}^p \lesssim \sum_{|l|\leq N} \|f * M_{\xi+\alpha l}\phi\|_{L^p}^p. \tag{3.10}$$

By raising estimate (3.10) to the q/p power, integrating over $[\alpha k, \alpha(k+1)]$, and using Lemma 1.14 we obtain

$$\|f * M_{\alpha k}\phi\|_{L^p}^q \lesssim \left(\sum_{|l|\leq N} \|f * M_{\xi+\alpha l}\phi\|_{L^p}^p \right)^{q/p}$$

$$\lesssim \int_{[\alpha k, \alpha(k+1)]} \left(\sum_{|l|\leq N} \|f * M_{\xi+\alpha l}\phi\|_{L^p}^p \right)^{q/p} d\xi$$

$$\lesssim \sum_{|l|\leq N} \int_{[\alpha k, \alpha(k+1)]} \|f * M_{\xi+\alpha l}\phi\|_{L^p}^q d\xi. \tag{3.11}$$

Finally, summing over $k \in \mathbb{Z}^d$ in (3.11) and then raising to the $1/q$ power give

$$|||f|||_{M^{p,q}} = \left(\sum_{k\in\mathbb{Z}^d} \|f*M_{\alpha k}\phi\|_{L^p}^q \right)^{1/q} \lesssim \left(\int_{\mathbb{R}^d} \|f * M_{\xi}\phi\|_{L^p}^q d\xi \right)^{1/q} = \|f\|_{M^{p,q}},$$

thus proving our second estimate.   □

The next result is a direct consequence of Theorem 3.4 and Proposition 1.59.

Theorem 3.5 *Let* $0 < p, q \leq \infty$, $\alpha, \beta > 0$ *be sufficiently small, and* $\phi \in \mathscr{S}_\alpha$ *be fixed. Then*

$$((f))_{M^{p,q}} := \left(\sum_{k\in\mathbb{Z}^d} \left(\sum_{l\in\mathbb{Z}^d} |(f * M_{\alpha k}\phi)(\beta l)|^p \right)^{q/p} \right)^{1/q} \tag{3.12}$$

is an equivalent quasi-norm on $M^{p,q}$*. When* $p = \infty$ *or* $q = \infty$*, the essential supremum is used.*

Note also that the quasi-norm in (3.12) can be rewritten as

$$((f))_{M^{p,q}} = \left\| \left((f * M_{\alpha k}\phi)(\beta l) \right)_{k,l \in \mathbb{Z}^d} \right\|_{\ell^p \ell^q}.$$

Any of the equivalent quasi-norms (3.1), (3.4), and (3.12) that define our modulation spaces depend so far on the choice of the window $\phi \in \mathscr{S}_\alpha$, where $\alpha > 0$ is sufficiently small. Our next result establishes that Definition 3.1 is, in fact, independent of the particular window ϕ.

Proposition 3.6 *Let* $0 < p, q \leq \infty$ *and* $\alpha > $ *be sufficiently small. The definition of* $M^{p,q}$ *is independent of the window* $\phi \in \mathscr{S}_\alpha$, *and different windows* $\phi_1, \phi_2 \in \mathscr{S}_\alpha$ *yield equivalent quasi-norms.*

Proof For a given f, we wish to show that we have the equivalence

$$\left\| (f * M_{\alpha k}\phi_1)_{k \in \mathbb{Z}^d} \right\|_{L^p \ell^q} \sim \left\| (f * M_{\alpha k}\phi_2)_{k \in \mathbb{Z}^d} \right\|_{L^p \ell^q}.$$

We borrow the notation and some of the ideas that yielded the first estimate in the proof of Theorem 3.4. Note first that, since $\phi_2 \in \mathscr{S}_\alpha$, we can find an $N = N(\alpha, d) \in \mathbb{N}$ such that

$$f * M_{\alpha k}\phi_1 = f * \left(T_{\alpha k}\widehat{\phi}_1 \sum_{|l| \leq N} T_{\alpha(k+l)}\widehat{\phi}_2 \right)^{\vee}$$

$$= f * \left(\sum_{|l| \leq N} M_{\alpha(k+l)}\phi_2 \right) * M_{\alpha k}\phi_1.$$

Mimicking now the argument that proves estimate (3.7), we have, for all $p > 0$, the following estimate

$$\left\| f * M_{\alpha k}\phi_1 \right\|_{L^p} \leq C \sum_{|l| \leq N} \left\| f * M_{\alpha(k+l)}\phi_2 \right\|_{L^p}, \tag{3.13}$$

where $C = \|\phi_1\|_{L^{\max(p,1)}}$. If we now raise (3.13) to the q-th power, use Lemma 1.14, and then sum over $k \in \mathbb{Z}^d$, we obtain

$$\sum_{k \in \mathbb{Z}^d} \left\| f * M_{\alpha k}\phi_1 \right\|_{L^p}^q \lesssim \sum_{k \in \mathbb{Z}^d} \sum_{|l| \leq N} \left\| f * M_{\alpha(k+l)}\phi_2 \right\|_{L^p}^q$$

$$\lesssim \sum_{k \in \mathbb{Z}^d} \left\| f * M_{\alpha k}\phi_2 \right\|_{L^p}^q.$$

Finally, raising the last estimate to the $1/q$ power yields $\left\| (f * M_{\alpha k}\phi_1)_{k \in \mathbb{Z}^d} \right\|_{L^p \ell^q} \lesssim \left\| (f * M_{\alpha k}\phi_2)_{k \in \mathbb{Z}^d} \right\|_{L^p \ell^q}$; the implicit constant in the right-hand side is dependent upon α, d, p and ϕ_1. Now simply reversing the roles of ϕ_1 and ϕ_2 concludes the proof. □

The remainder of the properties presented below have been already proved in Chap. 2 for the modulation spaces $\mathscr{M}^{p,q}$ with $1 \leq p, q \leq \infty$. The main point of our exposition here is to show that these basic properties hold also for positive indices below 1. For the sake of completeness, we will prove them for the whole range of indices $p, q \in (0, \infty]$; consequently, we also provide alternate proofs for some of the properties outlined in Chap. 2. Our immediate next result shows that the monotonicity (or nestedness) property of modulation spaces holds also for indices below 1.

Proposition 3.7 *If $0 < p_1 \leq p_2 \leq \infty$ and $0 < q_1 \leq q_2 \leq \infty$, then $M^{p_1,q_1} \subset M^{p_2,q_2}$.*

Proof Let $f \in M^{p_1,q_1}$, that is, for a fixed $\phi \in \mathscr{S}_\alpha$, with $\alpha > 0$ sufficiently small, we have

$$|||f|||_{M^{p_1,q_1}} = \|(f * M_{\alpha k}\phi)_{k \in \mathbb{Z}^d}\|_{L^{p_1}\ell^{q_1}} < \infty.$$

Fix $k \in \mathbb{Z}^d$. Because $\widehat{f * M_{\alpha k}\phi} = \widehat{f}(T_{\alpha k}\widehat{\phi})$, we see that supp $\widehat{f * M_{\alpha k}\phi} \subset \mathbb{B}(\alpha|k|, 1)$. By using now Proposition 1.57 we therefore obtain

$$\|f * M_{\alpha k}\phi\|_{L^{p_2}} \lesssim \|f * M_{\alpha k}\phi\|_{L^{p_1}}.$$

Finally, using the nestedness property of ℓ^p spaces (see Proposition 1.16), we obtain

$$\|(f * M_{\alpha k}\phi)_{k \in \mathbb{Z}^d}\|_{L^{p_2}\ell^{q_2}} \lesssim \|(f * M_{\alpha k}\phi)_{k \in \mathbb{Z}^d}\|_{L^{p_1}\ell^{q_1}},$$

thus proving $f \in M^{p_2,q_2}$ and $|||f|||_{M^{p_2,q_2}} \lesssim |||f|||_{M^{p_1,q_1}}$. $\square$

As for $\mathscr{M}^{p,q}$, we expect to have the space of rapidly decreasing Schwarz functions to be densely contained in $M^{p,q}$. The statement below confirms this expectation, see also [125, Proposition 11.3.4].

Proposition 3.8 *If $0 < p, q \leq \infty$, then $\mathscr{S}(\mathbb{R}^d)$ is a subspace of $M^{p,q}(\mathbb{R}^d)$ and $M^{p,q}(\mathbb{R}^d)$ is continuously embedded into $\mathscr{S}'(\mathbb{R}^d)$. Moreover, if $0 < p, q < \infty$, $\mathscr{S}(\mathbb{R}^d)$ is a dense subspace of $M^{p,q}$.*

Proof We begin by showing that $\mathscr{S}$ is a subspace of $M^{p,q}$ for all $0 < p, q \leq \infty$. Let $f \in \mathscr{S}$, $\phi \in \mathscr{S}_\alpha$, with $\alpha > 0$ sufficiently small, and $M, N \in \mathbb{N}$ be such that $M > d/p$ and $N > d/q$. By using integration by parts repeatedly and the product rule, we obtain the following sequence of equalities; integration by parts is justified because both $f, g \in \mathscr{S}$.

$$(f * M_\xi\phi)(x) = \int_{\mathbb{R}^d} f(x - y)\phi(y)e^{2\pi i\xi \cdot y}\, dy$$

$$= (1 + 4\pi^2|\xi|^2)^{-N/2} \int_{\mathbb{R}^d} f(x - y)\phi(y)(I - \Delta_y)^{N/2}(e^{2\pi i\xi \cdot y})\, dy$$

$$= (1 + 4\pi^2|\xi|^2)^{-N/2} \int_{\mathbb{R}^d} (I - \Delta_y)^{N/2}(f(x - y)\phi(y))e^{2\pi i\xi \cdot y}\, dy$$

$$= (1+4\pi^2|\xi|^2)^{-N/2} \sum_{|\alpha|\leq N} C_{N,\alpha} \int_{\mathbb{R}^d} \partial^\alpha f(x-y)\partial^{N-\alpha}\phi(y)e^{2\pi i\xi\cdot y}\, dy.$$

Since $1 + |x| \leq (1 + |x - y|)(1 + |y|)$ and $f, \phi \in \mathscr{S}$, the previous sequence of calculations yields the estimate

$$|(f * M_\xi\phi)(x)| \lesssim (1 + |x|)^{-M}(1 + |\xi|)^{-N} \sum_{|\alpha|\leq N} \|f\|_{M,\alpha};$$

for the definition of the seminorms of $\|\cdot\|_{M,\alpha}$, see again Sect. 1.2.2. The conditions $M > d/p$ and $N > d/q$ allow us now to integrate appropriately the previous estimate and obtain

$$\|(f * M_\xi\phi)(x)\|_{L_x^p L_\xi^q} \lesssim \sum_{|\alpha|\leq N} \|f\|_{M,\alpha} < \infty,$$

that is, $f \in M^{p,q}$ and $\|f\|_{M^{p,q}} \lesssim \sum_{|\alpha|\leq N} \|f\|_{M,\alpha}$. This proves that $\mathscr{S} \subset M^{p,q}$ for all $p, q \in (0, \infty]$.

To show that $M^{p,q}$ is continuously embedded in $\mathscr{S}'(\mathbb{R}^d)$ simply note that, because of Proposition 3.7, it suffices to show that $M^{\infty,\infty} \subset \mathscr{S}'$. Let then $f \in M^{\infty,\infty}$ and $\psi \in \mathscr{S}$. Clearly, $\psi \in M^{1,1}$, and for a sufficiently large $N \in \mathbb{N}$, we have

$$\|\psi\|_{M^{1,1}} \lesssim \sum_{|\alpha|\leq N} \|f\|_{N,\alpha}.$$

Since $\phi \in \mathscr{S}_\alpha$, there exists an integer $N = N(\operatorname{supp}\widehat{\phi}, \alpha, d) \in \mathbb{N}$ such that for all $k \in \mathbb{Z}^d$ we have $T_{\alpha k}\widehat{\phi} = \sum_{|l|\leq N} T_{\alpha(k+l)}\widehat{\phi}T_{\alpha k}\widehat{\phi}$. Recall also that $\sum_{k\in\mathbb{Z}^d} T_{\alpha k}\widehat{\phi} = 1$. Therefore,

$$|\langle f, \psi\rangle| = \left|\left\langle \sum_{k\in\mathbb{Z}^d}\left(\widehat{f}T_{\alpha k}\widehat{\phi}\sum_{|l|\leq N}T_{\alpha(k+l)}\widehat{\phi}\right)^\vee, \psi\right\rangle\right|$$

$$= \left|\sum_{k\in\mathbb{Z}^d}\sum_{|l|\leq N}\langle(f * M_{\alpha k}\phi) * M_{\alpha(k+l)}\phi, \psi\rangle\right|$$

$$= \left|\sum_{k\in\mathbb{Z}^d}\langle f * M_{\alpha k}\phi, \sum_{|l|\leq N}M_{\alpha(k+l)}\phi * \psi\rangle\right|$$

$$\leq \sum_{k\in\mathbb{Z}^d}\|f * M_{\alpha k}\phi\|_{L^\infty}\left\|\sum_{|l|\leq N}M_{\alpha(k+l)}\phi * \psi\right\|_{L^1}$$

$$\lesssim \|f\|_{M^{\infty,\infty}}\|\psi\|_{M^{1,1}}$$

$$\lesssim \|f\|_{M^{\infty,\infty}}\sum_{|\alpha|\leq N}\|f\|_{N,\alpha}.$$

We just proved that $f \in \mathscr{S}'$, and consequently $M^{\infty,\infty} \subset \mathscr{S}'$.

Let now $0 < p, q < \infty$; we wish to prove that $\mathscr{S}$ is dense in $M^{p,q}$. Fix for now $g \in M^{p,q} \cap \mathscr{C}^\infty$ and let $\varphi \in \mathscr{S}$ be a smooth bump with $\varphi(0) = 1$ and $\operatorname{supp}(\widehat{\varphi}) \subset \overline{\mathbb{B}(0,1)}$. For $n \geq 1$, let $g_n(x) = \varphi(n^{-1}x)g(x)$. Since $\varphi \in \mathscr{S}$, we immediately see that $g_n \in \mathscr{S}$, and

$$\|g_n - g\|_{M^{p,q}} \to 0 \text{ as } n \to \infty.$$

We have shown that any function in $M^{p,q} \cap \mathscr{C}^\infty$ can be approximated (in the $M^{p,q}$ norm) by a sequence of functions from $\mathscr{S}$. Thus, to finish our argument, it suffices to approximate (in the $M^{p,q}$ norm) any function in $M^{p,q}$ by a sequence of functions in $M^{p,q} \cap \mathscr{C}^\infty$. We achieve this as follows. Because $f \in M^{p,q} \subset \mathscr{S}'$ and $\phi \in \mathscr{S}$, we have $f * M_\xi \phi \in \mathscr{C}^\infty$ (in fact, tempered at infinity) for all $\xi \in \mathbb{R}^d$; thus

$$f_n = \sum_{|k| \geq n} f * M_{\alpha k}\phi \in M^{p,q} \cap \mathscr{C}^\infty$$

and, for some $N \in \mathbb{N}$,

$$\|f_n - f\|_{M^{p,q}} \lesssim \left(\sum_{|k| > n-N} \|f * M_{\alpha k}\phi\|_{L^p}^q \right)^{1/q} \to 0 \text{ as } n \to \infty.$$

$\square$

Our next result is concerned with the completeness of the modulation spaces.

Proposition 3.9 *If $0 < p, q \leq \infty$, then $M^{p,q}$ is a quasi-Banach space. If $1 \leq p, q \leq \infty$, then $M^{p,q}$ is a Banach space.*

Proof We only show the completeness of $M^{p,q}$. Let $\alpha, \beta > 0$ be sufficiently small, and $\phi \in \mathscr{S}_\alpha$ be fixed. By Theorem 3.5, we have the following equivalent quasi-norm on $M^{p,q}$:

$$((f))_{M^{p,q}} = \left\| \left((f * M_{\alpha k}\phi)(\beta l) \right)_{k,l \in \mathbb{Z}^d} \right\|_{\ell^p \ell^q}.$$

Write $c_{kl}(f) = (f * M_{\alpha k}\phi)(\beta l)$ and $\mathbf{c}(f) = (c_{kl}(f))_{k,l \in \mathbb{Z}^d}$. Note that $\mathbf{c} : M^{p,q} \to \mathbb{C}$ is a linear mapping; in particular, $\mathbf{c}(f) - \mathbf{c}(g) = \mathbf{c}(f - g)$. With this notation, we have

$$((f))_{M^{p,q}} = \|\mathbf{c}(f)\|_{\ell^{p,q}}.$$

Let $(f_n)_{n \geq 1}$ be a Cauchy sequence in $M^{p,q}$. Then $(\mathbf{c}(f_n))_{n \geq 1}$ is a Cauchy sequence in $\ell^{p,q}$. Since $\ell^{p,q}$ is a quasi-Banach space, there exists a sequence $\mathbf{c} = (c_{kl})_{k,l \in \mathbb{Z}^d} \in \ell^{p,q}$ such that

$$\lim_{n\to\infty} \|\mathbf{c}(f_n) - \mathbf{c}\|_{\ell^{p,q}} = 0.$$

Define now $f \in \mathscr{S}'$ so that $\langle f, T_{\beta l} M_{\alpha k}\phi \rangle = c_{kl}$, or equivalently, $\mathbf{c}(f) = \mathbf{c}$. This implies that $f \in M^{p,q}$ and $\|f_n - f\|_{M^{p,q}} \to 0$ as $n \to \infty$, which in turn proves the completeness of $M^{p,q}$.   □

Finally, let us state the duality of the modulation spaces. Let $0 < p < \infty$. If $p \geq 1$, we define the dual exponent p' in the usual way, that is, $1/p + 1/p' = 1$. When $p < 1$, we let $p' = \infty$.

Proposition 3.10 Let $0 < p, q < \infty$. Then $(M^{p,q})' = M^{p',q'}$.

A partial proof of this result is provided in [167]. In a subsequent paper, using the molecular characterization of modulation spaces, Kobayashi and Sawano [169] gave a complete proof of Proposition 3.10. In the next subsection, we will present a proof based on yet another equivalent quasi-norm of modulation space given in [155]. However, we observe that when $1 \leq p, q < \infty$, the proof follows from Theorem 2.6 and Proposition 3.2.

3.2 The Wiener Decomposition

Let $k \in \mathbb{Z}^d$. We denote by $\mathbb{Q}_k = k + [-1/2, 1/2]^d$ the unit cube in $\mathbb{R}^d$ centered at $k \in \mathbb{Z}^d$; in other words, $\mathbb{Q}_k = k + \mathbb{Q}_0$ is the translation by k of the unit cube in $\mathbb{R}^d$ centered at the origin.

Definition 3.11 The set $\{\mathbb{Q}_k : k \in \mathbb{Z}^d\}$ is called the Wiener decomposition of $\mathbb{R}^d$.

The Wiener decomposition of $\mathbb{R}^d$ induces a natural uniform decomposition in the frequency space of a signal via the (non-smooth) *frequency-uniform decomposition operators* $Q_k = \mathscr{F}^{-1}\chi_{\mathbb{Q}_k}\mathscr{F}$; note that, given some function f and for all $k \in \mathbb{Z}^d$, $Q_k(f)$ are uniformly localized in frequency. The drawback of this approach to decomposing the frequency content of a signal is the roughness of the characteristic functions $\chi_{\mathbb{Q}_k}$. However, we can fix this issue by replacing the characteristic functions by some appropriate smooth cut-off ones. Inspired by the $\mathscr{S}_\alpha$ notation used for the class of windows needed in the definition of the $M^{p,q}$ spaces of the previous subsection, we now introduce another good class of windows for our purposes; we call this class $\mathscr{S}_\star$.

Definition 3.12 We say that $\psi \in \mathscr{S}_\star(\mathbb{R}^d)$ if $\psi \in \mathscr{S}(\mathbb{R}^d)$, $\mathrm{supp}\,(\psi) \subset [-1, 1]^d$, $\psi(\xi) = 1$ for $\xi \in \mathbb{Q}_0$, $0 \leq \psi(\xi) \leq 1$ and $\sum_{k\in\mathbb{Z}^d} \psi(\xi - k) = 1$ for all $\xi \in \mathbb{R}^d$.

Let us now fix $\psi \in \mathscr{S}_\star(\mathbb{R}^d)$. We define next the smooth operators that are $\mathbb{Q}_k$ localized in frequency.

Definition 3.13 Denote by $\psi_k = T_k\psi = \psi(\cdot - k)$ the translation of ψ at k and let $P_k^\psi = \mathscr{F}^{-1}\psi_k\mathscr{F}$. We say that $\{P_k^\psi : k \in \mathbb{Z}^d\}$ is the set of *smooth frequency-uniform* *decomposition operators associated with* $\psi \in \mathscr{S}_\star$. When the (fixed) ψ is clear from the context, we will ignore the upper index of the decomposition operators and simply write P_k instead of P_k^ψ.

Written explicitly, the frequency-uniform decomposition operator P_k^ψ has the form

$$P_k^\psi(f)(x) = (\mathscr{F}^{-1}(\psi(\cdot - k)\mathscr{F}f(\cdot)))(x) = \int_{\mathbb{R}^d} \psi(\xi - k)\widehat{f}(\xi)e^{2\pi i x \cdot \xi}\, d\xi.$$
(3.14)

Note also that the translations ψ_k, of ψ by $k \in \mathbb{Z}^d$, satisfy the following properties:

(a) $|\psi_k(\xi)| \gtrsim 1$ for all $\xi \in Q_k$;
(b) $|\partial_\xi^\alpha \psi_k(\xi)| \le C_\alpha$ for all $\xi \in \mathbb{R}^d$ and $\alpha \in (\mathbb{N} \cup \{0\})^d$.

In a series of papers [136, 141, 154, 155] dealing with lowering the well-posedness threshold in several commonly studied equations of physics, Wang and his collaborators have introduced and studied the basic properties of some other, a priori different, modulation spaces; to distinguish them from the previously discussed formulations of modulation spaces, we will denote them here by $M_{p,q}$. The relevant observation that will transpire below is that they are again the same as Feichtinger's modulation spaces. For a detailed exposition on this topic from the point of view of Definition 3.14, see [142].

Definition 3.14 Let $0 < p, q \le \infty$, and let $\psi \in \mathscr{S}_\star(\mathbb{R}^d)$ be fixed. The modulation space $M_{p,q} = M_{p,q}(\mathbb{R}^d)$ is the set of all tempered distribution $f \in \mathscr{S}'(\mathbb{R}^d)$ for which

$$\|f\|_{M_{p,q}} := \left(\sum_{k \in \mathbb{Z}^d}\left(\int_{\mathbb{R}^d}\left|\int_{\mathbb{R}^d}\psi(\xi - k)\widehat{f}(\xi)e^{2\pi i x \cdot \xi}\, d\xi\right|^p dx\right)^{q/p}\right)^{1/q} < \infty.$$
(3.15)

When $p = \infty$ or $q = \infty$, the essential supremum is used.

The expression of the quasi-norm above can be further simplified into

$$\|f\|_{M_{p,q}} = \left(\sum_{k \in \mathbb{Z}^d}\|P_k(f)\|_{L^p}^q\right)^{1/q} = \|(P_k(f))_{k \in \mathbb{Z}^d}\|_{L^p \ell^q}.$$
(3.16)

Note that our definition of the $M_{p,q}$ modulation spaces depends so far on the choice of the window $\psi \in \mathscr{S}_\star$. As expected, Definition 3.14 is, in fact, independent of the particular window ψ we select.

Proposition 3.15 *Let $0 < p, q \le \infty$. The definition of $M_{p,q}$ is independent of the window $\psi \in \mathscr{S}_\star$, and different windows $\psi, \phi \in \mathscr{S}_\star$ yield equivalent quasi-norms.*

Proof We will write $\|f\|^{\psi}_{M_{p,q}}$ for the modulation norm of f which is induced by the frequency-uniform decomposition operators P^{ψ}_k. We will show that $\|f\|^{\psi}_{M_{p,q}} \lesssim \|f\|^{\phi}_{M_{p,q}}$. Clearly, by reversing the order of the windows ψ and ϕ, we will obtain $\|f\|^{\psi}_{M_{p,q}} \sim \|f\|^{\phi}_{M_{p,q}}$.

Following [136], let us write Λ for the set of all d-tuples $k \in \mathbb{Z}^d$ which satisfy $\overline{\mathbb{B}(k, d^{1/2})} \cap \overline{\mathbb{B}(0, d^{1/2})} \neq \emptyset$; Λ has at most $O(d^{1/2})$ elements. In particular, this also implies that

$$\|P^{\psi}_k(f)\|_{L^p} \lesssim \sum_{l \in \Lambda} \|P^{\phi}_{k+l}(P^{\psi}_k(f))\|_{L^p}. \tag{3.17}$$

Using now the identity

$$P^{\psi}_0(f) = M_k(P^{\psi}_{-k}(M_{-k}f)),$$

we see that

$$\|P^{\phi}_{k+l}(P^{\psi}_k(f))\|_{L^p} = \|P^{\psi_k \phi_{k+l}}_{-k}(M_{-k}f)\|_{L^p}. \tag{3.18}$$

Let $\mathbb{K} := \cup_{k \in \Lambda}\overline{\mathbb{B}(k, d^{1/2})}$; $\mathbb{K} \subset \mathbb{R}^d$ is a compact set. Using now Proposition 1.66 and estimate (3.18), for $s > s(p, d)$ we obtain

$$\|P^{\phi}_{k+l}(P^{\psi}_k(f))\|_{L^p} \lesssim \|\psi_k\|_{H^s}\|P^{\phi}_{k+l}(f)\|_{L^p}. \tag{3.19}$$

By construction, we know that $\|\psi_k\|_{H^s} \lesssim 1$ for all $k \in \mathbb{Z}^d$. Thus, (3.17) and (3.19) imply that

$$\|P^{\psi}_k(f)\|_{L^p} \lesssim \sum_{l \in \Lambda} \|P^{\phi}_{k+l}(f)\|_{L^p}.$$

Using now Lemma 1.14 (as explained several times in Sect. 3.1), the last inequality in turn proves that $\|f\|^{\psi}_{M_{p,q}} \lesssim \|f\|^{\phi}_{M_{p,q}}$. $\qquad\square$

Our next observation provides yet an equivalent norm on the space $M_{p,q}$, a fact that will be used shortly in identifying its dual space.

Proposition 3.16 Let $1 \leq p, q < \infty$. If $f \in M_{p,q}$, then, for a fixed $\psi \in \mathscr{S}_{\star}$, there exists $\mathbf{f} = (f_k) \in L^p \ell^q$ such that

$$f = \sum_{k \in \mathbb{Z}^d} P^{\psi}_k(f_k). \tag{3.20}$$

Moreover,

$$((f))_{M_{p,q}} := \inf\{\|\mathbf{f}\|_{L^p\ell^q} : f = \sum_{k\in\mathbb{Z}^d} P_k^\psi(f_k), \mathbf{f} = (f_k)_{k\in\mathbb{Z}^d} \in L^p\ell^q\} \qquad (3.21)$$

is an equivalent norm on $M_{p,q}$. *Conversely, if an* $f \in \mathscr{S}'$ *can be represented as in* (3.20) *such that the norm in* (3.21) *is finite, then* $f \in M_{p,q}$.

Proof Let $f \in M_{p,q}$. Using the notation in the proof of Proposition 3.15, let us write

$$f_k = \sum_{l\in\Lambda} P_{k+l}^\psi(f) \text{ and } \mathbf{f} = (f_k)_{k\in\mathbb{Z}^d}.$$

Clearly, $f = \sum_{k\in\mathbb{Z}^d} P_k(f_k)$ and $\|\mathbf{f}\|_{L^p\ell^q} \lesssim \|f\|_{M_{p,q}}$. Using now Bernstein's multiplier theorem (see Proposition 1.66), we see that

$$\|P_k(f)\|_{L^p} \lesssim \sum_{l\in\Lambda} \|P_k P_{k+l}(f_{k+l})\|_{L^p} \lesssim \sum_{l\in\Lambda} \|f_{k+l}\|_{L^p},$$

which immediately implies $\|f\|_{M_{p,q}} \lesssim \|\mathbf{f}\|_{L^p\ell^q}$. $\qquad\square$

Similar to Proposition 3.2, we can show that, for a fixed window in $\mathscr{S}_\star$ and for all $p, q > 0$, the modulation spaces $M_{p,q}$ are no different than $\mathscr{M}^{p,q}$. Indeed, the quasi-norm in (3.16) turns out to be equivalent to that on $\mathscr{M}^{p,q}$, see [155, Proposition 2.1].

Proposition 3.17 *Let* $0 < p, q \le \infty$ *and let* $\psi \in \mathscr{S}_\star$. *For all* $f \in \mathscr{S}'$ *we have* $\|f\|_{M_{p,q}}^\psi \sim \|f\|_{\mathscr{M}^{p,q}}$. *In particular,* $M_{p,q} = \mathscr{M}^{p,q}$ *for* $0 < p, q \le \infty$.

Proof Without loss of generality, let us assume $f \in \mathscr{S}$. Let us recall that the definition of $\mathscr{M}^{p,q}$ is independent of the selected nonzero window, call it ϕ, in $\mathscr{S}$. Thus we can select ϕ as we please; for our purposes, let us assume that $\widehat{\phi}$ is a smooth bump function compactly supported in $\mathbb{B}(0, R)$ with an $R = R(d)$ fixed such that $R \sim 10d^{1/2}$ and $\widehat{\phi}(\xi) = 1$ for $\xi \in \overline{\mathbb{B}(0, 3d^{1/2})}$. By the fundamental STFT identity (see Proposition 1.82) we know that

$$V_\phi f(x, \xi) = e^{-2\pi i x\cdot\xi} V_{\widehat{\phi}}\widehat{f}(\xi, -x) = e^{-2\pi i x\cdot\xi}(\mathscr{F}^{-1}\widehat{\phi}(\cdot - \xi)\mathscr{F}f)(x).$$

Using the mean value theorem, we will be able to find $\xi_k \in \mathbb{Q}_k$ such that

$$\|f\|_{\mathscr{M}^{p,q}}^\phi = \|V_\phi f(x, \xi)\|_{L^p,q} = \left(\int_{\mathbb{R}^d} \|\mathscr{F}^{-1}\widehat{\phi}(\cdot - \xi)\mathscr{F}f\|_{L^p}^q\right)^{1/q}$$

$$\sim \left(\sum_{k\in\mathbb{Z}^d} \|P_{\xi_k}^{\widehat{\phi}}(f)\|_{L^p}^q\right)^{1/q}. \qquad (3.22)$$

By the choice of ϕ and Bernstein's multiplier theorem (see Proposition 1.66), we have the following:

$$\|P_k(f)\|_{L^p} = \|\mathscr{F}^{-1}\psi_k\mathscr{F}f\|_{L^p} = \|\mathscr{F}^{-1}\psi_k\widehat{\phi}_{\xi_k}\mathscr{F}f\|_{L^p} \lesssim \|P^{\widehat{\phi}}_{\xi_k}(f)\|_{L^p}. \quad (3.23)$$

Using now (3.22), (3.23), and (3.16), we immediately get that $\|f\|_{M_{p,q}^{\psi}} \lesssim \|f\|_{\mathscr{M}^{p,q}}^{\phi}$. For the reverse inequality, let us observe that the support of $\widehat{\phi}_{\xi_k} = T_{\xi_k}\widehat{\phi}$ intersects at most $O(d^{1/2})$ many supports of $\psi_k = T_k\psi$. Using again Proposition 1.66, and recalling the notation for the set Λ of size $O(d^{1/2})$ utilized in the proof of Proposition 3.15, we obtain the following:

$$\|P^{\widehat{\phi}}_{\xi_k}(f)\|_{L^p} = \|\mathscr{F}^{-1}\sum_{l\in\Lambda}\psi_{k+l}\widehat{\phi}_{\xi_k}\mathscr{F}f\|_{L^p} \lesssim \sum_{l\in\Lambda}\|P^{\psi}_{k+l}(f)\|_{L^p}. \quad (3.24)$$

From (3.22), (3.24), and (3.16), we now get $\|f\|_{\mathscr{M}^{p,q}}^{\phi} \lesssim \|f\|_{M_{p,q}}^{\psi}$. This completes our proof.   □

Now that we observed the identifications $\mathscr{M}^{p,q} = M^{p,q} = M_{p,q}$ for all $0 < p, q \le \infty$, all the properties from Sect. 3.1 we proved for $M^{p,q}$ will also hold for the "new" modulation spaces $M_{p,q}$. For independent arguments proving these properties that use the frequency-uniform decomposition operators and (3.16), the reader can consult [136, 142, 155] and the relevant references therein. We finally turn our attention to Proposition 3.10; as promised, we will prove it in this section. Using the notation in Proposition 3.10, it suffices to show that $(M_{p,q})' = M_{p',q'}$ for $0 < p, q < \infty$. We will follow closely the argument from [155].

Proof of Proposition 3.10 We break the proof into several cases.

1. *The case* $1 \le p, q < \infty$. We show first that $M_{p',q'} \subset (M_{p,q})'$. Recall that $M_{p,q} \subset \mathscr{S}'$ for all $p, q > 0$. Let $g \in M_{p',q'}$, and $f \in \mathscr{S}$. First of all, note that we have the decomposition

$$\langle g, f\rangle = \sum_{k\in\mathbb{Z}^d}\sum_{l\in\Lambda}\langle P_{k+l}(g), P_k(f)\rangle.$$

From here, using the triangle inequality, and the continuous and discrete Cauchy-Schwartz inequalities, we get

$$|\langle g, f\rangle| \le \sum_{k\in\mathbb{Z}^d}\sum_{l\in\Lambda}\|P_{k+l}(g)\|_{L^{p'}}\|P_k(f)\|_{L^p} \lesssim \|g\|_{M_{p',q'}}\|f\|_{M_{p,q}}.$$

The density of $\mathscr{S}$ in $M_{p,q}$ shows, in fact, that the last displayed inequality holds for all $f \in M_{p,q}$, thus proving $M_{p',q'} \subset (M_{p,q})'$.

Second, we wish to prove the inclusion $(M_{p,q})' \subset M_{p',q'}$. By definition, see (3.16), the mapping $M_{p,q} \ni f \mapsto (P_k(f))_{k\in\mathbb{Z}^d} \in \mathscr{X} \subset L^p\ell^q$ is isometric from $M_{p,q}$ into the subspace $\mathscr{X}$ implicitly defined above. Therefore, if $F \in (M_{p,q})'$, we can think of F as belonging to $\mathscr{X}'$ and then extending F to a continuous functional in $(L^p \ell^q)'$. By some abuse of notation, we call this extended functional also F. By Proposition 1.18, for all $\mathbf{f} = (f_k)_{k \in \mathbb{Z}^d} \in L^p \ell^q$, we can represent F as

$$F(\mathbf{f}) = \sum_{k \in \mathbb{Z}^d} \int_{\mathbb{R}^d} g_k(x) f_k(x) \, dx,$$

where $\mathbf{g} = (g_k)_{k \in \mathbb{Z}^d} \in L^{p'} \ell^{q'}$ and $\|F\|_{(L^p \ell^q)'} = \|\mathbf{g}\|_{L^{p'} \ell^{q'}}$. Let now $f \in \mathscr{S}(\mathbb{R}^d)$. The inclusion $\mathscr{S} \subset M_{p,q}$ implies that $\tilde{\mathbf{f}} := (P_k(f))_{k \in \mathbb{Z}^d} \in L^p \ell^q$. Thus,

$$F(\tilde{\mathbf{f}}) = \sum_{k \in \mathbb{Z}^d} \int_{\mathbb{R}^d} g_k(x) P_k(f)(x) \, dx = \int_{\mathbb{R}^d} \sum_{k \in \mathbb{Z}^d} P_k(g_k)(x) f(x) \, dx.$$

Consequently, we have the identity $F = \sum_{k \in \mathbb{Z}^d} P_k(g_k)$ in a distributional sense. Using now Proposition 3.16, we see that

$$\|F\|_{M_{p',q'}} \lesssim \|\mathbf{g}\|_{L^{p'} \ell^{q'}} = \|F\|_{(L^p \ell^q)'},$$

thus proving the desired inclusion.

2. *The case $1 \leq p < \infty$ and $0 < q < 1$.* The inclusion $M_{p',\infty} \subset (M_{p,q})'$ is trivially implied by the embedding $M_{p,q} \subset M_{p,1}$. For the reverse inclusion, let $F \in (M_{p,q})' \subset \mathscr{S}'$ and $f \in \mathscr{S}$. Let us fix also $\psi \in \mathscr{S}_*$. Then, using $\langle \cdot, \cdot \rangle$ for the dual bracket, we have

$$|\langle P_k^\psi(F), f \rangle| \lesssim \|F\|_{(M_{p,q})'} \|P_k^\psi(f)\|_{M_{p,q}}.$$

From here, the usual trick using Bernstein's multiplier theorem gives

$$|\langle P_k^\psi(F), f \rangle| \lesssim \|F\|_{(M_{p,q})'} \sum_{l \in \Lambda} \|P_0^{\psi_k \psi_{k+l}}(f)\|_{L^p} \lesssim \|F\|_{(M_{p,q})'} \|f\|_{L^p}.$$

Thus, $(P_k(F))_{k \in \mathbb{Z}^d} \in L^{p'} \ell^\infty$, which is equivalent to $F \in M_{p',\infty}$.

3. *The case $0 < p, q < 1$.* We only need to show that $(M_{p,q})' \subset M_{\infty,\infty}$. Similar to the last case discussed, for $F \in (M_{p,q})'$ and $k \in \mathbb{Z}^d$ arbitrary, now we can write

$$\|P_k(F)\|_{L^\infty} = \sup_{x \in \mathbb{R}^d} |\langle F, \psi_k^\vee(x - \cdot) \rangle| \lesssim \|F\|_{(M_{p,q})'}.$$

Thus $(P_k(F))_{k \in \mathbb{Z}^d} \in L^\infty \ell^\infty$, which is equivalent to $F \in M_{\infty,\infty}$.

4. *The case $0 < p < 1$ and $1 \leq q < \infty$.* We only need to show that $(M_{p,q})' \subset M_{\infty,q'}$. Let $F \in (M_{p,q})'$ and $k \in \mathbb{Z}^d$ be arbitrary. Since $P_k(F) \in L^\infty$, we can

pick $x_k \in \mathbb{R}^d$ such that $\|P_k(F)\|_{L^\infty} \lesssim |P_k^\psi(F)(x_k)|$. Let $\mathbf{a} = (a_k)_{k \in \mathbb{Z}^d} \in \ell^q$, $\mathbf{F} = (\|P_k(F)\|_{L^\infty})_{k \in \mathbb{Z}^d}$, and define $f = \sum_{k \in \mathbb{Z}^d} a_k \psi_k^\vee(x_k - \cdot)$. We have

$$|\langle \mathbf{a}, \mathbf{F} \rangle| \leq \sum_{k \in \mathbb{Z}^d} |a_k| \|P_k(F)\|_{L^\infty} \lesssim |\langle F, f \rangle| \leq \|F\|_{(M_{p,q})'} \|f\|_{M_{p,q}} \lesssim \|F\|_{(M_{p,q})'} \|\mathbf{a}\|_{\ell^q}.$$

Thus, $\mathbf{F} \in \ell^{q'}$, that is, $(P_k(F))_{k \in \mathbb{Z}^d} \in L^\infty \ell^{q'}$, which in turn is equivalent to $F \in M_{\infty,q'}$. $\square$

3.3 Notes

It might seem idiosyncratic to wait until this chapter to present Feichtinger's original definition of the modulation spaces. However, we think that defining the modulation spaces via the STFT as done in Chap. 2 is a natural extension of Chap. 1, specifically Sect. 1.3, as our approach utilizes the tools developed there. Nonetheless, it is important to realize that both approaches illustrate the similarities between these spaces and the Besov and Triebel-Lizorkin spaces. Indeed, all these spaces are concrete realizations of Feichtinger and Gröchenig's theory of co-orbit spaces. To see this, one must recall that their theory can be carried out on LCA groups. In the case of the modulation spaces, the underlying group is the Heisenberg group, while in the case of the Besov and Triebel-Lizorkin spaces the said group is the affine group.

Chapter 4
Pseudodifferential Operators

The systematic study of linear pseudodifferential operators has drawn lots of its motivation from partial differential equations, quantum mechanics, and signal analysis. The pseudo-local character of these operators constituted one of the key properties that made them attractive to the researchers working in partial differential equations. Indeed, the pioneering works on this subject in the 1960s, as explored for example by Hörmander [151–153] and Kohn-Nirenberg [172], were guided by a deep connection to elliptic and hypo-elliptic equations. Importantly, the pseudo-locality of an underlying operator T yields precise statements about its wave front set, which can roughly be expressed in saying that the singularities of $T(f)$ can be found in the same regions as those of f. Moreover, boundedness results on the classical spaces of harmonic analysis play a special role due to their implications in the regularity of the solutions for the appropriate equations; see also [218]. In quantum mechanics, pseudodifferential operators T_σ appear as quantizations of the symbol σ and they attempt to capture rigorously an observable of both the position and momentum operators simultaneously. In this context, the estimates on the symbol σ aim to respect as faithfully as possible the Heisenberg principle while also being sufficiently strong for establishing a meaningful theory. Finally, in signal analysis the so-called spreading representation of T_σ as a superposition of time-frequency shifts is useful in the modeling of time-varying systems and communication channels. Due to the nature of the modulation spaces, this last point of view will be essential to our purposes.

Multilinear (m-linear) pseudodifferential operators are natural generalizations of a product of m functions and they are equally relevant in the study of certain nonlinear partial differential equations. The work of Coifman and Meyer [54–58] introduced them as relevant tools in the study of nonlinear operators. They appear in the form of fractional Leibniz rules or Kato-Ponce inequalities in the study of the Euler and Navier-Stokes equations [159], the Kortweg-de Vries equation [51], as well as in the smoothing properties of Schrödinger semigroups [133]. These Leibniz-type estimates can be generalized to the context of multilinear

© Springer Science+Business Media, LLC, part of Springer Nature 2020
Á. Bényi, K. A. Okoudjou, *Modulation Spaces*, Applied and Numerical
Harmonic Analysis, https://doi.org/10.1007/978-1-0716-0332-1_4

pseudodifferential operators with Hörmander symbols, which in turn can be viewed, for appropriate parameters, as one of the facets of the theory of multilinear Calderón-Zygmund operators proposed by Grafakos-Torres [122]. A lot of the push for a comprehensive theory of multilinear pseudodifferential operators as we know it today owes to Calderón's work [36, 37] in the early 1970s on the Cauchy integral on Lipschitz curves and the related first order Calderón commutator. This lead to a search for uniform estimates on the so-called bilinear Hilbert transform that was eventually settled by Lacey-Thiele [174, 175] and Grafakos-Li [121]. Interestingly, the bilinear Hilbert transform ties in naturally with the Calderón-Zygmund method of rotations [44], only now performed bilinearly. Moreover, and more importantly, these investigations on Calderón's conjecture spurred the development of powerful time-frequency decomposition techniques reminiscent of the works of Carleson [41] and Fefferman [92] on the almost everywhere convergence of Fourier series.

In this chapter we will mainly be concerned with finding natural time-frequency conditions on the symbol of a pseudodifferential operator that imply the continuity of the corresponding operator on appropriate modulation spaces. Our presentation has the unifying feature of treating linear and multilinear pseudodifferential operators simultaneously, with appropriate distinctions being made as needed.

4.1 Translation Invariant Operators

Let $k \in \mathbb{N}$ and $L : \mathscr{S}(\mathbb{R}^d) \times \mathscr{S}(\mathbb{R}^d) \times \cdots \times \mathscr{S}(\mathbb{R}^d) \to \mathscr{S}'(\mathbb{R}^d)$ be a k-*linear operator*, that is, linear in each of its entries. When $k = 1$, we simply call L to be linear, when $k = 2$ we refer to such an L as a bilinear operator, when $k = 3$ as trilinear, and so on; in general, for $k \geq 2$, we refer to such objects as *multilinear operators*.

Definition 4.1 We say that L is *translation invariant* if for all $f_i \in \mathscr{S}$, $1 \leq i \leq k$, and for all $h \in \mathbb{R}^d$ we have

$$T_h(L(f_1, f_2, \ldots, f_k)) = L(T_h f_1, T_h f_2, \ldots, T_h f_k).$$

Example 4.2 Let $m \in \mathbb{N}$. If L is a (linear) partial differential operator with constant coefficients or order m,

$$L = \sum_{|\alpha| \leq m} c_\alpha \partial^\alpha,$$

we check easily that $T_h(L(f)) = L(T_h f)$ for all $h \in \mathbb{R}^d$ and $f \in \mathscr{S}$, thus L is translation invariant.

Similarly, for $m_j \in \mathbb{N}$, $1 \leq j \leq k$, $m = \sum_{j=1}^k m_j$, we can talk about a k-linear partial differential operator with constant coefficients of order m,

$$L(f_1, f_2, \ldots, f_k) = \sum_{j=1}^{k} \sum_{|\alpha_j| \leq m_j} c_{\alpha_1, \alpha_2, \ldots, \alpha_k} \prod_{j=1}^{k} \partial^{\alpha_j} f_j.$$

It is straightforward to check that such an L is translation invariant. In particular, the product of k functions is an example of a translation invariant k-linear operator.

It is well known, see for example [119], that a translation invariant linear operator L which maps L^p into L^q for some $1 \leq p, q \leq \infty$ is a convolution operator; that is, there exists a tempered distribution $u \in \mathscr{S}'(\mathbb{R}^d)$ such that $L(f) = u * f$ for all $f \in \mathscr{S}$. By taking the Fourier transform on both sides of this equality, we obtain $\widehat{L(f)} = \hat{u}\hat{f}$; that is, a translation invariant linear operator is simply a multiplication operator on the frequency side. Now, by applying the inverse Fourier transform, the previous equality becomes $L(f)(x) = \langle \hat{u}(\xi), \widehat{\hat{f}(\xi)e^{-2\pi i x \cdot \xi}} \rangle$. On a formal level, letting $\hat{u} = \sigma$, we thus have $L(e^{2\pi i x \cdot \xi}) = \sigma(\xi)e^{2\pi i x \cdot \xi}$ for all $\xi \in \mathbb{R}^d$. Now, further assuming that *the symbol σ is a locally integrable function which is tempered at infinity*, we can write

$$L(f)(x) = \int_{\mathbb{R}^d} \sigma(\xi)\hat{f}(\xi)e^{2\pi i x \cdot \xi}\, d\xi.$$

To further make explicit the dependence of the operator L on the symbol σ, we will write L_σ instead of L. In particular, by Plancherel's identity (see Proposition 1.51), L can be extended to a bounded operator on L^2 if and only if $\sigma \in L^\infty$.

An analogous discussion can be made in the multilinear setting. If the k-linear operator L is translation invariant and it maps $L^{p_1} \times L^{p_2} \times \cdots \times L^{p_k}$ into L^p for some $1 \leq p, p_j \leq \infty, 1 \leq j \leq k$, then there exists a tempered distribution u on $\mathbb{R}^{kd}$ such that

$$T(f_1, f_2, \ldots, f_k)(x) = (u * (f_1 \otimes f_2 \otimes f_k))(x, x, \ldots, x)$$

for all $f_j \in \mathscr{S}(\mathbb{R}^d), 1 \leq j \leq k$, where $(f_1 \otimes f_2 \otimes f_k)(x_1, x_2, \ldots, x_k) = \prod_{j=1}^{k} f_j(x_j)$. Denoting again by σ the Fourier transform of u (in $\mathbb{R}^{kd}$), we obtain

$$L_\sigma(f_1, f_2, \ldots, f_k)(x) =$$

$$\int_{\mathbb{R}^{kd}} \sigma(\xi_1, \xi_2, \ldots, \xi_k)\hat{f}_1(\xi_1)\hat{f}_2(\xi_2) \cdots \hat{f}_k(\xi_k)e^{2\pi i x \cdot (\xi_1 + \xi_2 + \cdots + \xi_k)}\, d\xi_1 d\xi_2 \cdots d\xi_k.$$

4.2   Linear and Multilinear Pseudodifferential Operators

Pseudodifferential operators are natural generalizations of the translation invariant operators. They are similarly characterized by a symbol σ, but now this symbol is allowed to depend on both the time variable x and the frequency variable ξ, so that

$$L_\sigma(e^{2\pi i x \cdot \xi}) = \sigma(x, \xi)e^{2\pi i x \cdot \xi}.$$

Suppose that $\sigma \in \mathscr{S}(\mathbb{R}^{2d})$ and define the operator L_σ on $\mathscr{S}(\mathbb{R}^d)$ by

$$L_\sigma(f)(x) = \int_{\mathbb{R}^d} \sigma(x, \xi)\widehat{f}(\xi)e^{2\pi i x \cdot \xi}\, d\xi. \tag{4.1}$$

It follows that L_σ defines a linear mapping from $\mathscr{S}$ to $\mathscr{S}$. Furthermore, we can extend it as a mapping from $\mathscr{S}$ to $\mathscr{S}'$ via the duality relation

$$\langle L_\sigma(f), g \rangle = \langle \sigma, W_1(g, f) \rangle, \tag{4.2}$$

where, for $f, g \in \mathscr{S}$,

$$W_1(g, f)(x, \xi) = g(x)\overline{\widehat{f}(\xi)}e^{-2\pi i x \cdot \xi}$$

denotes the *cross-Ryhaczek distribution* of f and g. Consequently, we arrive at the following definition.

Definition 4.3 Let $\sigma \in \mathscr{S}'(\mathbb{R}^{2d})$. The operator L_σ given by (4.1) is called the (linear) *pseudodifferential operator with symbol* σ. The mapping $\mathscr{S}'(\mathbb{R}^{2d}) \ni \sigma \mapsto L_\sigma \in \mathscr{S}'(\mathbb{R}^d)$ is called the *Kohn-Nirenberg correspondence*.

An alternate notation used in the literature for the pseudodifferential operator L_σ is $\sigma(x, D)$; in the sequel, we prefer to use the former notation.

Example 4.4 Let $\sigma(x, \xi) = \sum_{|\alpha| \leq m} c_\alpha(x)(2\pi i \xi)^\alpha$. Then, recalling that the derivative operator is nothing but a multiplication on the frequency side (see Proposition 1.46), we easily see that L_σ is a partial differential operator with variable coefficients of order m,

$$L_\sigma(f)(x) = \sum_{|\alpha| \leq m} c_\alpha(x)\partial^\alpha f(x).$$

In particular, if the symbol $\sigma(x, \xi) = \sigma(x)$ is independent of the frequency variable ξ, then $L_\sigma(f) = \sigma f$, thus L_σ is simply the multiplication operator by σ.

Example 4.5 If $\sigma(x, \xi) = \sigma(\xi)$ is independent of the spatial variable x, then, using Proposition 1.46 (g), we get $L_\sigma(f) = \sigma^\vee * f$. In this case, L_σ is commonly referred to as a *Fourier multiplier operator.*

Example 4.6 Let now $\sigma(x, \xi) = (\sigma_1 \otimes \sigma_2)(x, \xi) := \sigma_1(x)\sigma_2(\xi)$. Examples 4.4 and 4.5 tell us then that, in this case, the linear pseudodifferential operator L_σ is a *product-convolution operator*, $L_\sigma(f) = \sigma_1(\sigma_2^\vee * f)$.

A multilinear pseudodifferential operator L_σ satisfies the formal equation

$$L_\sigma(e^{2\pi i x \cdot \xi_1}, e^{2\pi i x \cdot \xi_2}, \ldots, e^{2\pi i x \cdot \xi_k}) = \sigma(x, \xi_1, \xi_2, \ldots, \xi_k)e^{2\pi i x \cdot (\xi_1 + \xi_2 + \cdots + \xi_k)}.$$

Definition 4.7 Let $k \in \mathbb{N}$ and $\sigma \in \mathscr{S}'(\mathbb{R}^{(k+1)d})$. The operator $L_\sigma : \mathscr{S}(\mathbb{R}^d) \times \mathscr{S}(\mathbb{R}^d) \times \cdots \times \mathscr{S}(\mathbb{R}^d)$ into $\mathscr{S}'(\mathbb{R}^d)$ defined by

$$L_\sigma(f_1, f_2, \ldots, f_k)(x) = \tag{4.3}$$

$$\int_{\mathbb{R}^{kd}} \sigma(x, \xi_1, \xi_2, \ldots, \xi_k) \widehat{f_1}(\xi_1) \widehat{f_2}(\xi_2) \cdots \widehat{f_k}(\xi_k) e^{2\pi i x \cdot (\xi_1 + \xi_2 + \cdots + \xi_k)} \, d\xi_1 d\xi_2 \cdots d\xi_k$$

is called the (k-linear) *pseudodifferential operator with symbol σ*.

As before, in the extension to $\mathscr{S}'$ we use the duality relation

$$\langle L_\sigma(f_1, f_2, \ldots, f_k), g \rangle = \langle \sigma, W_k(g, f_1, f_2, \ldots, f_k) \rangle, \tag{4.4}$$

where, for $f_j, g \in \mathscr{S}$, $1 \leq j \leq k$, we denoted

$$W_k(g, f_1, f_2, \ldots, f_k)(x, \xi_1, \xi_2, \ldots, \xi_k) = g(x) e^{-2\pi i x \cdot (\xi_1 + \xi_2 + \cdots + \xi_k)} \prod_{j=1}^{k} \widehat{f_j}(\xi),$$

the multilinear version of the cross-Ryhaczek distribution W_1 defined above.

Remark 4.8 Taking $k = 1$ in Definition 4.7 leads to the classical (linear) pseudodifferential operators given in Definition 4.3. Furthermore, one can also consider the case $k = 0$ in which case the symbol $\sigma \in \mathscr{S}'(\mathbb{R}^d)$ and the corresponding operator L_σ is simply a Fourier multiplier. Consequently, in the sequel, the context will make it clear when we are dealing with multilinear pseudodifferential operator $k \geq 2$, a linear pseudodifferential operator $k = 1$, or a Fourier multiplier $k = 0$.

Example 4.9 Let $\sigma(x, \xi_1, \xi_2, \ldots, \xi_k) = \sum_{j=1}^{k} \sum_{|\alpha_j| \leq m_j} c_{\alpha_1, \alpha_2, \ldots, \alpha_k}(x) \prod_{j=1}^{k} (2\pi i \xi_j)^{\alpha_j}$. As in Example 4.4, we see that L_σ is a k-linear partial differential operator with variable coefficients of order $m = \sum_{j=1}^{k} m_j$,

$$L_\sigma(f_1, f_2, \ldots, f_k) = \sum_{j=1}^{k} \sum_{|\alpha_j| \leq m_j} c_{\alpha_1, \alpha_2, \ldots, \alpha_k}(x) \prod_{j=1}^{k} \partial^{\alpha_j} f_j.$$

Example 4.10 Let us consider the convolution operator $H_1(f) = \text{p.v.} \frac{1}{x} * f$. This operator is known as the *Hilbert transform* and it plays an important role in harmonic analysis, where it is a first example of a singular integral operator [40]. H_1 is a (linear) pseudodifferential operator L_σ with x-independent symbol (or multiplier) $\sigma(\xi) = -i\pi \, \text{sign}(\xi)$, that is

$$H_1(f)(x) = \int_{\mathbb{R}} \sigma(\xi) \widehat{f}(\xi) e^{2\pi i x \cdot \xi} \, d\xi.$$

Example 4.11 Consider now the *bilinear Hilbert transform*

$$H_2(f_1, f_2)(x) = \text{p.v.} \int_{\mathbb{R}} f_1(x - t) f_2(x + t) \frac{dt}{t}.$$

This is a difficult object to study and, as already alluded to above, the developments surrounding the modern theory of "hard" time-frequency analysis owe a lot to it; a treatment of H_2 in the realm of modulation spaces can be found in [185]. A straightforward calculation shows that $H_2 = L_\sigma$ with bilinear multiplier $\sigma(\xi_1, \xi_2) = c\,\text{sign}(\xi_1 - \xi_2)$, that is

$$H_2(f_1, f_2)(x) = \int_{\mathbb{R}^2} \sigma(\xi_1, \xi_2) \widehat{f_1}(\xi_1) \widehat{f_2}(\xi_2) e^{2\pi i x \cdot (\xi_1 + \xi_2)} \, d\xi_1 d\xi_2.$$

In the following, we will use the notations

$$\vec{\xi} = (\xi_1, \xi_2, \ldots, \xi_k), \ d\vec{\xi} = d\xi_1 d\xi_2 \cdots d\xi_k,$$

$$\vec{f} = (f_1, f_2, \ldots, f_k), \ \vec{\alpha} = (\alpha_1, \alpha_2, \ldots, \alpha_k),$$

and

$$\partial_{\vec{\xi}}^{\vec{\alpha}} = \partial_{\xi_1}^{\alpha_1} \partial_{\xi_2}^{\alpha_2} \cdots \partial_{\xi_k}^{\alpha_k}.$$

In this context, we also write $\| \vec{\xi} \| = \sum_{j=1}^k |\xi_j|$ and $\| \vec{\alpha} \| = \sum_{j=1}^k |\alpha_j|$; note that α_j are themselves multi-indices and $|\alpha_j|$ has a meaning, see Sect. 1.2.2.

4.3   The Hörmander Classes of Symbols

Let us now return to the partial differential operator L_σ in Example 4.4 having as its symbol the characteristic polynomial $\sigma(x, \xi) = \sum_{|\alpha| \leq m} c_\alpha(x)(2\pi i \xi)^\alpha$. If we further assume that all coefficient functions $c_\alpha \in \mathscr{C}^\infty(\mathbb{R}^d)$ and all partial derivatives are bounded, $|\partial_x^\beta c_\alpha(x)| \leq C_{\alpha,\beta}$ for all $x \in \mathbb{R}^d$, we can easily check that

$$|\partial_x^\beta \partial_\xi^\gamma \sigma(x, \xi)| \leq C_{\beta,\gamma}(1 + |\xi|)^{m - |\gamma|}$$

for all multi-indices β, γ and for all $(x, \xi) \in \mathbb{R}^{2d}$. This and other examples arising in PDEs, such as the symbols coming from the Cauchy-Szegö projection operator or the approximate inverse for the heat equation, motivate the definition of the *classical Hörmander classes of symbols* $S_{\rho,\delta}^m$; here, $\rho, \delta, m \in \mathbb{R}$.

Definition 4.12 Let $\sigma \in \mathscr{C}^\infty(\mathbb{R}^{2d})$. We say that $\sigma \in S^m_{\rho,\delta}$ if for all $x, \xi \in \mathbb{R}^d$ and all multi-indices α, β we have

$$|\partial_x^\alpha \partial_\xi^\beta \sigma(x, \xi)| \lesssim (1 + |\xi|)^{m+\delta|\alpha|-\rho|\beta|}. \tag{4.5}$$

With this definition, the characteristic polynomial of a linear partial differential operator of order m belongs to the symbol class $S^m_{1,0}$. Similar considerations lead us to the *multilinear Hörmander classes of symbols*.

Definition 4.13 Let $\sigma \in \mathscr{C}^\infty(\mathbb{R}^{(k+1)d})$. We say that $\sigma \in S^m_{\rho,\delta}(k)$ if for all $(x, \vec{\xi}) \in \mathbb{R}^d \times \mathbb{R}^{kd}$ and all $\alpha, \vec{\beta}$ we have

$$|\partial_x^\alpha \partial_{\vec{\xi}}^{\vec{\beta}} \sigma(x, \vec{\xi})| \lesssim (1 + \|\vec{\xi}\|)^{m+\delta|\alpha|-\rho\|\vec{\beta}\|}. \tag{4.6}$$

Clearly, when $k = 1$, we simply recover the classes $S^m_{\rho,\delta} := S^m_{\rho,\delta}(1)$, while when $k = 2$, we obtain the bilinear classes of symbols $BS^m_{\rho,\delta} := S^m_{\rho,\delta}(2)$ introduced in [15]. Recalling now the multilinear symbol in Example 4.9, if we further assume there that the coefficients are all smooth and with all partial derivatives bounded, we see again that the characteristic polynomial of a k-linear partial differential operator of order m belongs to $S^m_{1,0}(k)$. The differential inequalities (4.6) can be used to construct in a natural way a family of seminorms on $S^m_{\rho,\delta}(k)$,

$$p^{m,\rho,\delta}_{\alpha,\vec{\beta}}(\sigma) = \sup\{(1+\|\vec{\xi}\|)^{-m-\delta|\alpha|+\rho\|\vec{\beta}\|}|\partial_x^\alpha \partial_{\vec{\xi}}^{\vec{\beta}} \sigma(x, \vec{\xi})| : (x, \vec{\xi}) \in \mathbb{R}^d \times \mathbb{R}^{kd}\},$$

that turn all these classes of symbols into Fréchet spaces.

Nowadays, many of the properties of pseudodifferential operators with Hörmander symbols are well understood. For instance, we have available a symbolic calculus and have a good understanding of boundedness properties in the scale of Lebesgue, Sobolev, or Besov spaces for the classes of symbols $S^m_{\rho,\delta}(k)$ with $0 \leq \delta \leq \rho \leq 1$. Let us briefly discuss the linear case $k = 1$. In [39], Calderón and Vaillancourt proved that $S^0_{\delta,\delta}, 0 \leq \delta < 1$, and hence also the smaller classes $S^0_{1,\delta}, 0 \leq \delta < 1$, yield bounded pseudodifferential operators on $L^2 = \mathscr{M}^{2,2}$; they achieved this by reducing matters to the L^2 boundedness of the class $S^0_{0,0}$, which in turn was proved in [38]. Moreover, the multilinear Hörmander classes of symbols $S^m_{1,\delta}(k)$, where $0 \leq \delta < 1$ and $m \in \mathbb{R}$ behave similarly well in the classical functional setting. For a nice summary of results pertaining to the linear setting, see Stein's book [208, Chapters 6 and 7]; for the bilinear case, see, for example, [7, 10, 15, 19, 21, 22, 181, 183].

The modulation spaces, by contrast, are ill suited for the study of the Hörmander classes of symbols $S^m_{1,\delta}$ with $0 \leq \delta < 1$ and $m \geq 0$; the special case $\mathscr{M}^{2,2}$ is the only exception. The following result is due to Sugimoto and Tomita [212].

Theorem 4.14 *Let* $1 < p, q < \infty$, $m \in \mathbb{R}$, *and* $0 < \delta < 1$. *If* $m > -\delta d|1/q - 1/2|$, *then there exists a symbol* $\sigma \in S_{1,\delta}^m$ *such that* L_σ *is not bounded on* $\mathscr{M}^{p,q}(\mathbb{R}^d)$. *In particular, the classes* $S_{1,\delta}^0$ *yield unbounded pseudodifferential operators on* $\mathscr{M}^{p,q}$, $q \neq 2$.

Closely related to Theorem 4.14 is another result of Sugimoto-Tomita [213].

Theorem 4.15 *Let* $1 < p, q < \infty$, $m \in \mathbb{R}$, $0 \leq \delta \leq \rho \leq 1$, $\delta < 1$, *and let* $\sigma \in S_{\rho,\delta}^m$. *There exists a "critical index"* $m(p,q) \leq 0$ *such that if* $m \leq m(p,q)$, *then* L_σ *is bounded on* $\mathscr{M}^{p,q}(\mathbb{R}^d)$. *In particular,* L_σ *is bounded on* $\mathscr{M}^{2,q}$ *if and only if* $m \leq -\delta d|1/q - 1/2|$.

For a better understanding of Theorems 4.14 and 4.15, let us consider the broader spectrum of Calderón-Zygmund operators. It can be shown that Calderón-Zygmund operators of convolution type are always bounded on the modulation spaces. A relevant manifestation of this fact is the $\mathscr{M}^{p,q}$ boundedness of the Hilbert transform H_1 (see Example 4.10), firstly observed in [17]. But the boundedness on modulation spaces fails, in general, for non-convolution type Calderón-Zygmund operators, even if we impose additional reasonable conditions implying boundedness on other classical function spaces; see [213].

Let us now return to the $S_{0,0}^0$ class of symbols. Note that, for $p \neq 2$, this class fails, in general, to yield bounded pseudodifferential operators on L^p. Moreover, in bilinear setting, even the expected $L^2 \times L^2 \rightarrow L^1$ boundedness fails, as shown in [16]. In fact, the $L^p \times L^q \rightarrow L^r$ boundedness of the class $BS_{0,0}^0$ fails, in general, for all exponents satisfying the Hölder condition $1/p + 1/q = 1/r$; see, for example, [22]. While the classical functional setting cannot capture the characteristics of the classes $S_{0,0}^0(k)$, the realm of modulation spaces can. As we shall see in the next section, the modulation spaces $\mathscr{M}^{\infty,1}$ and, more generally, $\mathscr{M}^{p,1}$, $1 \leq p \leq \infty$, are good symbol classes for the time-frequency analysis of multilinear pseudodifferential operators generalizing the Calderón-Vaillancourt ones of order zero. Moreover, modulation spaces will now play the dual role of both symbol classes and spaces of functions on which the pseudodifferential operators act.

Remark 4.16 In the next section, we will prove some boundedness results for k-linear pseudodifferential operators, with arbitrary $k \in \mathbb{N}$. This is convenient because, in particular, we recover the corresponding results for linear pseudodifferential operators by simply letting $k = 1$. We wish to point out that the converse is not true, that is, one cannot infer the boundedness of a k-linear pseudodifferential operator defined on $\mathbb{R}^d$ simply from the boundedness of a linear pseudodifferential operator defined on $\mathbb{R}^{kd}$.

For ease of notation, let us consider the case $k = 2$. Suppose that L_σ is a linear pseudodifferential operator on $\mathbb{R}^{2d}$, that is, for appropriate functions F on $\mathbb{R}^{2d}$, we have

$$L_\sigma(F)(\overrightarrow{x}) = \int_{\mathbb{R}^{2d}} \sigma(\overrightarrow{x}, \overrightarrow{\xi}) e^{2\pi i \overrightarrow{x} \cdot \overrightarrow{\xi}} \widehat{F}(\overrightarrow{\xi}) \, d\overrightarrow{\xi}. \tag{4.7}$$

Here, we use the notation introduced before: $\overrightarrow{x} = (x_1, x_2)$ and $\overrightarrow{\xi} = (\xi_1, \xi_2)$ are both in $\mathbb{R}^{2d}$. Now, if F is defined through a tensor, that is,

$$F(\overrightarrow{x}) = f_1(x_1) f_2(x_2),$$

for some f_1, f_2 defined on $\mathbb{R}^d$, then (4.7) can be rewritten as

$$L_\sigma(\overrightarrow{f})(x_1, x_2) = \int_{\mathbb{R}^{2d}} \sigma(x_1, x_2, \overrightarrow{\xi}) e^{2\pi i (x_1 \cdot \xi_1 + x_2 \cdot \xi_2)} \widehat{f_1}(\xi_1) \widehat{f_2}(\xi_2) \, d\overrightarrow{\xi}. \tag{4.8}$$

Thus, the restriction of the linear pseudodifferential operator L_σ to the diagonal $\{(x, x) : x \in \mathbb{R}^d\}$ yields a bilinear pseudodifferential operator L_τ, with symbol τ defined by

$$\tau(x, \overrightarrow{\xi}) := \sigma(x, x, \overrightarrow{\xi}).$$

More precisely, we have

$$L_\tau(\overrightarrow{f})(x) = \int_{\mathbb{R}^{2d}} \sigma(x, x, \overrightarrow{\xi}) e^{2\pi i x \cdot (\xi_1 + \xi_2)} \widehat{f_1}(\xi_1) \widehat{f_2}(\xi_2) \, d\overrightarrow{\xi}.$$

For the sake of the argument, let us assume that the linear L_σ maps $L^p(\mathbb{R}^{2d})$ into $L^q(\mathbb{R}^{2d})$ for some p, q, that is

$$\|L_\sigma(F)\|_{L^q(\mathbb{R}^{2d})} \lesssim \|F\|_{L^p(\mathbb{R}^{2d})}; \tag{4.9}$$

the argument for modulation spaces would follow a similar pattern.

But, for a tensor like function F as above, we would then have, via Fubini's theorem, that $\|F\|_{L^p(\mathbb{R}^{2d})} = \|f_1\|_{L^p(\mathbb{R}^d)} \|f_2\|_{L^p(\mathbb{R}^d)}$, yet clearly the $L^q(\mathbb{R}^{2d})$ norm of $L_\sigma(F)$ cannot be related to the $L^q(\mathbb{R}^d)$ norm of L_τ. Consequently, the boundedness in (4.9) provides no information, say, about the $L^p(\mathbb{R}^d) \times L^p(\mathbb{R}^d) \to L^q(\mathbb{R}^d)$ boundedness of L_τ simply because Lebesgue integration disregards sets of measure zero. Moreover, note that, due to the pointwise nature of the estimates, if one starts with the linear symbol $\sigma \in S^m_{\rho,\delta}$ on $\mathbb{R}^{2d}$, then the bilinear symbol τ (on $\mathbb{R}^d$) will belong to $BS^m_{\rho,\delta}$.

4.4   Modulation Spaces as Classes of Symbols

We begin by providing a brief motivation for the natural appearance of modulation spaces as symbol classes for the linear pseudodifferential operators defined in (4.1). Specifically, we show that we can express

$$L_\sigma(f)(x) = \int_{\mathbb{R}^d} \sigma(x, \xi) \widehat{f}(\xi) e^{2\pi i x \cdot \xi} \, d\xi$$

as a superposition of time-frequency shifts with the *spreading function* $\widehat{\sigma}$,

$$L_\sigma(f)(x) = \int_{\mathbb{R}^d} \int_{\mathbb{R}^d} \widehat{\sigma}(\eta, y)(M_\eta T_{-y} f)(x) \, dy \, d\eta. \tag{4.10}$$

This follows from the following formal sequence of equalities; here, $\mathscr{F}_1$ and $\mathscr{F}_2$ denote the Fourier transform in the first and second variables, respectively.

$$
\begin{aligned}
L_\sigma(f)(x) &= \int_{\mathbb{R}^d} \sigma(x, \xi) \left(\int_{\mathbb{R}^d} f(y) e^{-2\pi i \xi \cdot y} \, dy \right) e^{2\pi i x \cdot \xi} \, d\xi \\
&= \int_{\mathbb{R}^d} \left(\int_{\mathbb{R}^d} \sigma(x, \xi) e^{-2\pi i (y-x) \cdot \xi} \, d\xi \right) f(y) \, dy \\
&= \int_{\mathbb{R}^d} \mathscr{F}_2 \sigma(x, y - x) f(y) \, dy = \int_{\mathbb{R}^d} \mathscr{F}_1^{-1} \widehat{\sigma}(x, y - x) f(y) \, dy \\
&= \int_{\mathbb{R}^d} \left(\int_{\mathbb{R}^d} \widehat{\sigma}(\eta, y - x) e^{2\pi i \eta \cdot x} \, d\eta \right) f(y) \, dy \\
&= \int_{\mathbb{R}^d} \int_{\mathbb{R}^d} \widehat{\sigma}(\eta, y) e^{2\pi i \eta \cdot x} f(x + y) \, d\eta \, dy \\
&= \int_{\mathbb{R}^d} \int_{\mathbb{R}^d} \widehat{\sigma}(\eta, y)(M_\eta T_{-y} f)(x) \, dy \, d\eta.
\end{aligned}
$$

In light of (4.10), the study of $L_\sigma(f)$ then becomes a question about the interaction between the symbol σ and the time-frequency concentration of the function or distribution f that the pseudodifferential operator is acting upon. We thus arrive to the modulation spaces, whose norms effectively capture the information provided by the time-frequency shifts.

Let us now return to the most general set-up of multilinear pseudodifferential operators. Recall that, by (4.4), the action of such a k-linear operator on a $k+1$ tuple of functions $(g, \overrightarrow{f}) = (g, f_1, f_2, \dots, f_k) \in \mathscr{S}(\mathbb{R}^d) \times \mathscr{S}(\mathbb{R}^d) \times \cdots \times \mathscr{S}(\mathbb{R}^d)$ can be written in terms of the multilinear cross-Rychaczek distribution W_k as

$$\langle L_\sigma(\overrightarrow{f}), g \rangle = \langle \sigma, W_k(g, \overrightarrow{f}) \rangle.$$

Letting now $(\phi_0, \overrightarrow{\phi}) = (\phi_0, \phi_1, \phi_2, \ldots, \phi_k) \in \mathscr{S}(\mathbb{R}^{(k+1)d})$, we clearly have $W_k(\phi_0, \overrightarrow{\phi}) \in \mathscr{S}(\mathbb{R}^{(k+1)d})$. Thus, using Parseval's STFT identity, see Proposition 1.82, we can further express this identity as

$$\langle L_\sigma(\overrightarrow{f}), g \rangle = \langle V_{W_k(\phi_0, \overrightarrow{\phi})}\sigma, V_{W_k(\phi_0, \overrightarrow{\phi})}W_k(g, \overrightarrow{f})\rangle. \tag{4.11}$$

This formula suggests that the analysis of multilinear pseudodifferential operators on modulation spaces is rooted in the understanding of W_k. This intuition turns out to be correct, and it is further supported by the following "magic formula" proved in [18]; for a similar formula in linear setting ($k = 1$) in which the cross-Rychaczek distribution W_1 is replaced by the cross-Wigner distribution, see [125, Lemma 14.5.1].

Lemma 4.17 *Let* $(g, \overrightarrow{f}) \in (\mathscr{M}^{\infty,\infty}(\mathbb{R}^d))^{k+1}$, *and* $(u_0, \overrightarrow{u}) = (u_0, u_1, \ldots, u_k)$, $(v_0, \overrightarrow{v}) = (v_0, v_1, \ldots, v_k) \in \mathbb{R}^d \times \mathbb{R}^{kd}$. *We have*

$$V_{W_k(\phi_0, \overrightarrow{\phi})}W_k(g, \overrightarrow{f})((u_0, \overrightarrow{u}), (v_0, \overrightarrow{v}))$$

$$= e^{-2\pi i \sum_{j=1}^k u_j \cdot v_j} V_{\phi_0}g(u_0, v_0 + \sum_{j=1}^k u_j) \prod_{j=1}^k \overline{V_{\phi_j}f_j(u_0 + v_j, u_j)}. \tag{4.12}$$

Proof We show first that (4.12) holds if $(g, \overrightarrow{f}) \in (\mathscr{S}(\mathbb{R}^d))^{k+1}$. Since $W_k(\phi_0, \overrightarrow{\phi}) \in \mathscr{S}(\mathbb{R}^{(k+1)d})$, the integral defining $V_{W_k(\phi_0, \overrightarrow{\phi})}W_k(g, \overrightarrow{f})$ converges absolutely; hence the following manipulations are justified.

$$V_{W_k(\phi_0, \overrightarrow{\phi})}W_k(g, \overrightarrow{f})((u_0, \overrightarrow{u}), (v_0, \overrightarrow{v}))$$

$$= \int_{\mathbb{R}^{(k+1)d}} W_k(g, \overrightarrow{f})(x, \overrightarrow{\xi}) e^{-2\pi i (x, \overrightarrow{\xi}) \cdot (v_0, \overrightarrow{v})} \overline{W_k(\phi_0, \overrightarrow{\phi})((x, \overrightarrow{\xi}) - (u_0, \overrightarrow{u}))} \, dx \, d\overrightarrow{\xi}$$

$$= \int_{\mathbb{R}^{(k+1)d}} g(x) \prod_{j=1}^k \overline{\widehat{f_j}(\xi_j)} e^{-2\pi i x \cdot \sum_{j=1}^k \xi_j} e^{-2\pi i (x \cdot v_0 + \sum_{j=1}^k \xi_j \cdot v_j)} \times$$

$$\overline{\phi_0(x - u_0)} \prod_{j=1}^k \overline{\widehat{\phi_j}(\xi_j - u_j)} e^{2\pi i (x - u_0) \cdot \sum_{j=1}^k (\xi_j - u_j)} \, dx \, d\overrightarrow{\xi}$$

$$= e^{2\pi i u_0 \cdot \sum_{j=1}^k u_j} \int_{\mathbb{R}^{(k+1)d}} g(x) \prod_{j=1}^k \overline{\widehat{f_j}(\xi_j)} \, \overline{\phi_0(x - u_0)} \prod_{j=1}^k \overline{\widehat{\phi_j}(\xi_j - u_j)} \times$$

$$e^{-2\pi i x \cdot (v_0 + \sum_{j=1}^k u_j)} \prod_{j=1}^k e^{-2\pi i \xi_j \cdot (u_0 + v_j)} \, dx \, d\vec{\xi}$$

$$= e^{2\pi i u_0 \cdot \sum_{j=1}^k u_j} \, V_{\phi_0} g\left(u_0, v_0 + \sum_{j=1}^k u_j\right) \prod_{j=1}^k \overline{V_{\widehat{\phi_j}} \widehat{f_j}(u_j, -u_0 - v_j)}$$

$$= e^{2\pi i u_0 \cdot \sum_{j=1}^k u_j} \, V_{\phi_0} g\left(u_0, v_0 + \sum_{j=1}^k u_j\right) \prod_{j=1}^k \overline{V_{\phi_j} f_j(u_0 + v_j, u_j)} e^{-2\pi i \sum_{j=1}^k u_j \cdot (u_0 + v_j)}$$

$$= e^{-2\pi i \sum_{j=1}^k u_j \cdot v_j} \, V_{\phi_0} g(u_0, v_0 + \sum_{j=1}^k u_j) \prod_{j=1}^k \overline{V_{\phi_j} f_j(u_0 + v_j, u_j)}.$$

The general case, when $(g, \vec{f}) \in (\mathcal{M}^{\infty,\infty}(\mathbb{R}^d))^{k+1}$, is obtained by approximation. The three relevant facts needed in this step are:

(1) $\mathcal{M}^{\infty,\infty}$ is left invariant by a "chirp" multiplication, see the remark immediately following Theorem 12.1.3 in [125];
(2) $\mathcal{S}$ is weak*-dense in $\mathcal{M}^{\infty,\infty}$;
(3) Weak*-convergence of tempered distributions is equivalent to uniform convergence of the STFT on compact sets, see [100].

Now, if $(g, \vec{f}) \in (\mathcal{M}^{\infty,\infty}(\mathbb{R}^d))^{k+1}$, from Fact (1) we get that $W_k(g, \vec{f}) \in \mathcal{M}^{\infty,\infty}$, thus $V_{W_k(\phi_0, \vec{\phi})} W_k(g, \vec{f})$ is a bounded and uniformly continuous function on $\mathbb{R}^{2(k+1)d}$. By Fact (2), let $(g_n, \vec{f_n}) \in (\mathcal{S}(\mathbb{R}^d))^{k+1}$ such that $(g_n, \vec{f_n}) \overset{w^*}{\to} (g, \vec{f})$ in $\mathcal{M}^{\infty,\infty}$. Because tensor products and multiplication by chirps are continuous operations, we get $W_m(g_n, \vec{f_n}) \overset{w^*}{\to} W_m(g, \vec{f})$ in $\mathcal{M}^{\infty,\infty}$. Finally, by Fact (3), $V_{W_m(\phi_0, \vec{\phi})} W_m(g_n, \vec{f_n})$ converges uniformly on compact sets to $V_{W_m(\phi_0, \vec{\phi})} W_m(g, \vec{f})$. A similar reasoning then proves that $e^{-2\pi i \sum_{j=1}^k u_j \cdot v_j} V_{\phi_0} g_n(u_0, v_0 + \sum_{j=1}^k u_j) \prod_{j=1}^k \overline{V_{\phi_j} (f_n)_j(u_0 + v_j, u_j)}$ converges uniformly to $e^{-2\pi i \sum_{j=1}^k u_j \cdot v_j} V_{\phi_0} g(u_0, v_0 + \sum_{j=1}^k u_j) \prod_{j=1}^k \overline{V_{\phi_j} f_j(u_0 + v_j, u_j)}$, which shows that (4.12) holds in this general case. $\qquad\Box$

Our next result states a mixed Lebesgue estimate for the STFT of $W_k(g, \vec{f})$.

Proposition 4.18 *Let* $1 \leq p, p_j, q_j \leq \infty, 0 \leq j \leq k$, *be such that*

$$\frac{1}{p_0'} + \sum_{j=1}^k \frac{1}{p_j} = \frac{1}{p'} \quad and \quad \frac{1}{q_0'} + \sum_{j=1}^k \frac{1}{q_j} = \frac{k}{p'}.$$

If $f_j \in \mathcal{M}^{p_j, q_j}$ and $g \in \mathcal{M}^{p'_0, q'_0}$, then $W_k(g, \vec{f}) \in \mathcal{M}^{p', \infty}$; moreover,

$$\| V_{W_k(\phi_0, \vec{\phi})} W_k(g, \vec{f}) \|_{L^{p'}, \infty} \lesssim \| g \|_{\mathcal{M}^{p'_0, q'_0}} \prod_{j=1}^{k} \| f_j \|_{\mathcal{M}^{p_j, q_j}}.$$

Proof Let us denote by

$$\Delta(v_0, \vec{v}) = \left(\int_{\mathbb{R}^{(k+1)d}} |V_{W_k(\phi_0, \vec{\phi})} W_k(g, \vec{f})((u_0, \vec{u}), (v_0, \vec{v}))|^{p'} \, du_0 \, d\vec{u} \right)^{1/p'}.$$

By Lemma 4.17, we have

$$\Delta(v_0, \vec{v})^{p'} = \int_{\mathbb{R}^{(k+1)d}} |V_{\phi_0} g(u_0, v_0 + \sum_{j=1}^{k} u_j)|^{p'} \left| \prod_{j=1}^{k} V_{\phi_j} f_j (u_0 + v_j, u_j) \right|^{p'} du_0 \, d\vec{u}.$$

Now, using the first condition on the p_j exponents and Hölder's inequality (1.2) in the variable u_0, we obtain

$$\Delta(v_0, \vec{v})^{p'} \leq \int_{\mathbb{R}^{kd}} \left\| V_{\phi_0} g(\cdot, v_0 + \sum_{j=1}^{k} u_j) \right\|_{L^{p'_0}}^{p'} \prod_{j=1}^{k} \| V_{\phi_j} f_j(\cdot, u_j) \|_{L^{p_j}}^{p'} d\vec{u}.$$

Let $G(v) = \| V_{\phi_0} g(\cdot, v) \|_{L^{p'_0}}$ and $F_j(u_j) = \| V_{\phi_j} f_j(\cdot, -u_j) \|_{L^{p_j}}$; with this notation, $\| g \|_{\mathcal{M}^{p'_0, q'_0}} = \| G \|_{L^{q'_0}}$ and $\| f_j \|_{\mathcal{M}^{p_j, q_j}} = \| F_j \|_{L^{q_j}}$. Moreover, the last inequality can now be restated as

$$\Delta(v_0, \vec{v})^{p'} \leq (G^{p'} * F_1^{p'} * \cdots * F_k^{p'})(v_0).$$

Using now the second condition on the q_j exponents and Young's convolution inequality (1.8), we obtain

$$\| V_{W_k(\phi_0, \vec{\phi})} W_k(g, \vec{f}) \|_{L^{p'}, \infty} = \| \Delta(v_0, \vec{v}) \|_{L^{\infty}(dv_0 d\vec{v})} = \| \Delta(v_0, \vec{v})^{p'} \|_{L^{\infty}(dv_0 d\vec{v})}^{1/p'}$$

$$\leq \| G^{p'} * F_1^{p'} * \cdots * F_k^{p'} \|_{L^{\infty}(dv_0)}^{1/p'}$$

$$\leq \left(\| G^{p'} \|_{L^{q'_0/p'}} \prod_{j=1}^{k} \| F_j^{p'} \|_{L^{q_j/p'}} \right)^{1/p'}$$

$$= \| G \|_{L^{q'_0}} \prod_{j=1}^{k} \| F_j \|_{L^{q_j}} \lesssim \| g \|_{\mathcal{M}^{p'_0, q'_0}} \prod_{j=1}^{k} \| f_j \|_{\mathcal{M}^{p_j, q_j}}.$$

Note that in the very last estimate, the implicit constant depends only on the dimension d and the indices of the modulation spaces and it is implied by the equivalence of the modulation space norm under different windows.   □

The identity (4.11) and Proposition 4.18 allow us to prove a boundedness result about k-linear pseudodifferential operators with symbols in the modulation class $\mathcal{M}^{p,1}(\mathbb{R}^{(k+1)d})$, $1 \le p \le \infty$; see again [12].

Theorem 4.19 *Let $\sigma \in \mathcal{M}^{p,1}(\mathbb{R}^{(k+1)d})$ and $1 \le p, p_j, q_j \le \infty$, $0 \le j \le k$, be such that*

$$\frac{1}{p_0'} + \sum_{j=1}^{k} \frac{1}{p_j} = \frac{1}{p'} \quad and \quad \frac{1}{q_0'} + \sum_{j=1}^{k} \frac{1}{q_j} = \frac{k}{p'}. \tag{4.13}$$

Then the k-linear pseudodifferential operator L_σ defined by (4.3) can be extended to a bounded operator from $\mathcal{M}^{p_1,q_1} \times \mathcal{M}^{p_2,q_2} \times \cdots \times \mathcal{M}^{p_k,q_k}$ into $\mathcal{M}^{p_0,q_0}$. Moreover, the following estimate holds:

$$\|L_\sigma(\overrightarrow{f})\|_{\mathcal{M}^{p_0,q_0}} \lesssim \|\sigma\|_{\mathcal{M}^{p,1}} \prod_{j=1}^{k} \|f_j\|_{\mathcal{M}^{p_j,q_j}}. \tag{4.14}$$

Proof We will use, in this order, the identity (4.11), Hölder's inequality for mixed Lebesgue spaces (1.6), and Proposition 4.18. For any $g \in \mathcal{M}^{p_0',q_0'}$ and $f_j \in \mathcal{M}^{p_j,q_j}$, $1 \le j \le k$, we have

$$|\langle L_\sigma(\overrightarrow{f}), g\rangle| = |\langle V_{W_k(\phi_0, \overrightarrow{\phi})}\sigma, V_{W_k(\phi_0, \overrightarrow{\phi})}W_k(g, \overrightarrow{f})\rangle|$$

$$\le \|V_{W_k(\phi_0, \overrightarrow{\phi})}\sigma\|_{L^{p,1}} \|V_{W_k(\phi_0, \overrightarrow{\phi})}W_k(g, \overrightarrow{f})\|_{L^{p',\infty}}$$

$$\lesssim \|\sigma\|_{\mathcal{M}^{p,1}} \prod_{j=1}^{k} \|f_j\|_{\mathcal{M}^{p_j,q_j}} \|g\|_{\mathcal{M}^{p_0',q_0'}}.$$

Now, if $p_0' < \infty$ and $q_0' < \infty$, then the duality properties of the modulation spaces imply that $L_\sigma(\overrightarrow{f}) \in \mathcal{M}^{p_0,q_0}$ with the norm estimate (4.14). To deal with the cases where either $p_0' = \infty$ or $q_0' = \infty$ or both, we recall that the related modulation spaces satisfy similar duality relations, that is, $(\mathcal{M}^{0,q_0})' = \mathcal{M}^{1,q_0'}$, $(\mathcal{M}^{p_0,0})' = \mathcal{M}^{p_0',1}$, and $(\mathcal{M}^{0,0})' = \mathcal{M}^{1,1}$. The stated identifications of dual spaces allow us to run the same duality argument in these cases as well and conclude our proof.   □

For the remainder of this section, we will explore some of the consequences of Theorem 4.19 and their connections with recent and not so recent results about linear and multilinear pseudodifferential operators on modulation spaces.

If we let $p = 1$ in Theorem 4.19, we necessarily have $p_0 = q_0 = 1$ and $p_j = q_j = \infty$ for $j = 1, \ldots, k$. Since $\mathscr{M}^{1,1} \subset \mathscr{M}^{p,q} \subset \mathscr{M}^{\infty,\infty}$ and $\mathscr{M}^{1,1} \subset L^p \subset \mathscr{M}^{\infty,\infty}$ for all $1 \le p, q \le \infty$, we thus obtain the following result; see [12, Remark 5].

Corollary 4.20 *If $\sigma \in \mathscr{M}^{1,1}(\mathbb{R}^{(k+1)d})$, then L_σ is bounded from $\mathscr{M}^{\infty,\infty} \times \mathscr{M}^{\infty,\infty} \times \cdots \times \mathscr{M}^{\infty,\infty}$ into $\mathscr{M}^{1,1}$. In particular, if $\sigma \in \mathscr{M}^{1,1}(\mathbb{R}^{(k+1)d})$, then L_σ is bounded from $\mathscr{M}^{p_1,q_1} \times \mathscr{M}^{p_2,q_2} \times \cdots \times \mathscr{M}^{p_k,q_k}$ into $\mathscr{M}^{p_0,q_0}$, and from $L^{p_1} \times L^{p_2} \times \cdots \times L^{p_k}$ into L^{p_0} for all $1 \le p_j, q_j, p_0, q_0 \le \infty, 1 \le j \le k$.*

If we let $p = \infty$ in Theorem 4.19, we recover the main result from [18, Theorem 3.1] about the boundedness of L_σ with σ in $\mathscr{M}^{\infty,1}$-the so-called *Sjöstrand class*, see [206].

Corollary 4.21 *Let $\sigma \in \mathscr{M}^{\infty,1}(\mathbb{R}^{(k+1)d})$ and $1 \le p_j, q_j \le \infty, 0 \le j \le k$, be such that*

$$\sum_{j=1}^{k} \frac{1}{p_j} = \frac{1}{p_0} \quad \text{and} \quad \sum_{j=1}^{k} \frac{1}{q_j} = k - 1 + \frac{1}{q_0}. \tag{4.15}$$

Then the k-linear pseudodifferential operator L_σ defined by (4.3) can be extended to a bounded operator from $\mathscr{M}^{p_1,q_1} \times \mathscr{M}^{p_2,q_2} \times \cdots \times \mathscr{M}^{p_k,q_k}$ into $\mathscr{M}^{p_0,q_0}$. In particular, the same boundedness holds if $\sigma \in S_{0,0}^0(k)$.

By letting $\sigma = 1$ in Corollary 4.21, and with the same exponents as there, leads to the multilinear estimate

$$\left\| \prod_{i=1}^{k} u_i \right\|_{\mathscr{M}^{p_0,q_0}} \lesssim \prod_{i=1}^{k} \|u_i\|_{\mathscr{M}^{p_i,q_i}}. \tag{4.16}$$

Intuitively, the previous corollary states that the modulation symbol class $\mathscr{M}^{\infty,1}$ yields multilinear pseudodifferential operators that behave like pointwise multiplication in both the spatial and frequency variables. Indeed, the index relations expressed in (4.15) are precisely the conditions that are required to apply Hölder's inequality (1.2) to the product $\prod_{j=1}^{k} f_j$ and the Young inequality (1.8) to the k-fold convolution $\widehat{f_1} * \widehat{f_2} * \cdots \widehat{f_k}$.

Furthermore, by letting $p_j = q_j$ in Corollary 4.21, we necessarily obtain that either $k = 1$ and $p_0 = q_0$, or $k = 2$, $p_0 = 1$, and $q_0 = \infty$. The first situation recovers the classical result of Gröchenig and Heil [127, Theorem 1.1], which in turn extends a classical result of Calderón and Vaillancourt about linear pseudodifferential operators with symbols in $S_{0,0}^0$. The second situation is an extension of [18, Corollary 4.4]. In particular, it provides the correct set-up for dealing with bilinear symbols of Calderón-Vaillancourt type which are known not to be bounded on Lebesgue spaces; see again the discussion preceding this section which is rooted in the results from [16].

Corollary 4.22 *If $\sigma \in \mathscr{M}^{\infty,1}(\mathbb{R}^{2d})$ and $1 \leq r \leq \infty$, then the linear pseudodifferential operator L_σ can be extended to a bounded operator from $\mathscr{M}^{r,r}$ into $\mathscr{M}^{r,r}$. In particular, if $\sigma \in S_{0,0}^0$, then L_σ can be extended to a bounded operator from $\mathscr{M}^{r,r}$ into $\mathscr{M}^{r,r}$, and hence from L^2 into L^2.*

The proof of the second part of Corollary 4.22 is based on the embedding $S_{0,0}^0(\mathbb{R}^{2d}) \subset \mathscr{M}^{\infty,1}(\mathbb{R}^{2d})$ [125, Theorem 14.5.3]. In fact, this is a simple consequence of a more subtle embedding $\mathscr{C}^s(\mathbb{R}^{2d}) \subset \mathscr{M}^{\infty,1}(\mathbb{R}^{2d})$ for $s > 2d$ which was established by Heil et al. [147, Proposition 6.3]. Here, $\mathscr{C}^s(\mathbb{R}^{2d})$ is the Hölder-Lipschitz space of order s [208]. It is worth pointing out that the modulation space $\mathscr{M}^{\infty,1}(\mathbb{R}^d)$ was also independently introduced by Sjöstrand who proved that it was an algebra of pseudodifferential operators larger than those generated by the Calderón-Vaillancourt class $S_{0,0}^0$ [206]. Other early and independent investigations of the properties of $\mathscr{M}^{\infty,1}(\mathbb{R}^d)$ are due to Boulkhemair [35].

Corollary 4.23 *Let $\sigma \in \mathscr{M}^{\infty,1}(\mathbb{R}^{3d})$ and $1 \leq r \leq \infty$. Then the bilinear pseudodifferential operator L_σ can be extended to a bounded operator from $\mathscr{M}^{r,r} \times \mathscr{M}^{r',r'}$ into $\mathscr{M}^{1,\infty}$. In particular, if $\sigma \in BS_{0,0}^0$, then L_σ can be extended to a bounded operator from $\mathscr{M}^{r,r} \times \mathscr{M}^{r',r'}$ into $\mathscr{M}^{1,\infty}$, and hence from $L^2 \times L^2$ into $\mathscr{M}^{1,\infty} \supset L^1$.*

If we now choose $q_j = p'_j$ in Corollary 4.21, we recover [18, Corollary 4.2].

Corollary 4.24 *Let $\sigma \in \mathscr{M}^{\infty,1}(\mathbb{R}^{(k+1)d})$ and $1 \leq p_j \leq \infty$, $0 \leq j \leq k$, be such that*

$$\sum_{j=1}^{k} \frac{1}{p_j} = \frac{1}{p_0}.$$

Then the k-linear pseudodifferential operator L_σ can be extended to a bounded operator from $\mathscr{M}^{p_1,p'_1} \times \mathscr{M}^{p_2,p'_2} \times \cdots \times \mathscr{M}^{p_k,p'_k}$ into $\mathscr{M}^{p_0,p'_0}$. In particular, the same boundedness holds if $\sigma \in S_{0,0}^0(k)$.

In keeping with the previous analysis, letting then $k = 1$, respectively $k = 2$, in Corollary 4.24, we obtain the following results that resemble those of Corollaries 4.22 and 4.23.

Corollary 4.25 *If $\sigma \in \mathscr{M}^{\infty,1}(\mathbb{R}^{2d})$ and $1 \leq r \leq \infty$, then the linear pseudodifferential operator L_σ can be extended to a bounded operator from $\mathscr{M}^{r,r'}$ into $\mathscr{M}^{r,r'}$. In particular, if $\sigma \in S_{0,0}^0$, then L_σ can be extended to a bounded operator from $\mathscr{M}^{r,r'}$ into $\mathscr{M}^{r,r'}$, and hence from L^2 into L^2.*

Corollary 4.26 *Let $\sigma \in \mathscr{M}^{\infty,1}(\mathbb{R}^{3d})$ and $1 \leq p_j \leq \infty$, $0 \leq j \leq 2$, be such that $1/p_1 + 1/p_2 = 1/p_0$. Then the bilinear pseudodifferential operator L_σ can be extended to a bounded operator from $\mathscr{M}^{p_1,p'_1} \times \mathscr{M}^{p_2,p'_2}$ into $\mathscr{M}^{p_0,p'_0}$. In particular,* *if* $\sigma \in BS_{0,0}^0$, *then* L_σ *can be extended to a bounded operator from* $\mathcal{M}^{p_1,p_1'} \times \mathcal{M}^{p_2,p_2'}$ *into* $\mathcal{M}^{p_0,p_0'}$, *and hence from* $L^2 \times L^2$ *into* $\mathcal{M}^{1,\infty}$.

Let us now turn back our attention to the generic modulation symbol class $\mathcal{M}^{p,1}$, $1 < p < \infty$. As before, we wish to explore the consequences of our main result, Theorem 4.19, when we impose some special relations on the indices of the spaces p_j, q_j.

Letting $p_j = q_j$, $1 \le j \le k$, we obtain the following result.

Corollary 4.27 *Let* $\sigma \in \mathcal{M}^{p,1}(\mathbb{R}^{(k+1)d})$ *and* $1 \le p, p_j \le \infty$, $0 \le j \le k$, *be such that*

$$\sum_{j=1}^k \frac{1}{p_j} = \frac{1}{p_0} - \frac{1}{p} \quad and \quad \frac{1}{q_0} = \frac{1}{p_0} - \frac{k-1}{p'}. \tag{4.17}$$

Then the k-linear pseudodifferential operator L_σ *can be extended to a bounded operator from* $\mathcal{M}^{p_1,p_1} \times \mathcal{M}^{p_2,p_2} \times \cdots \times \mathcal{M}^{p_k,p_k}$ *into* $\mathcal{M}^{p_0,q_0}$.

We note immediately that (4.17) is imposing an upper bound on the integer $k \ge 1$. For example, we must trivially have $k \le 1 + p'$. Thus, assuming that the symbol $\sigma \in \mathcal{M}^{p,1}(\mathbb{R}^{(k+1)d})$ with $p \ge 2$, then necessarily $k \le 3$; in other words, we can talk at most of a trilinear pseudodifferential operator in this case. For the benefit of the reader, let us consider the specific situation in which $p = 2$ and $k = 3$. It is easy to see then that, necessarily, $p_0 = 1$ and $q_0 = \infty$; thus, Corollary 4.27 would read in this case as follows.

Corollary 4.28 *Let* $\sigma \in \mathcal{M}^{2,1}(\mathbb{R}^{4d})$ *and* $1 \le p_j \le \infty$, $1 \le j \le 3$, *be such that*

$$\frac{1}{p_1} + \frac{1}{p_2} + \frac{1}{p_3} = \frac{1}{2}.$$

Then the trilinear pseudodifferential operator L_σ *can be extended to a bounded operator from* $\mathcal{M}^{p_1,p_1} \times \mathcal{M}^{p_2,p_2} \times \mathcal{M}^{p_3,p_3}$ *into* $\mathcal{M}^{1,\infty}$. *In particular,* L_σ *is bounded from* $\mathcal{M}^{6,6} \times \mathcal{M}^{6,6} \times \mathcal{M}^{6,6}$ *into* $\mathcal{M}^{1,\infty}$.

Letting now $p_j' = q_j$, $1 \le j \le k$, in Theorem 4.19, we arrive at the following statement.

Corollary 4.29 *Let* $\sigma \in \mathcal{M}^{p,1}(\mathbb{R}^{(k+1)d})$ *and* $1 \le p, p_j \le \infty$, $0 \le j \le k$, *be such that*

$$\sum_{j=1}^k \frac{1}{p_j} = \frac{1}{p_0} - \frac{1}{p} \quad and \quad \frac{1}{q_0} = \frac{1}{p_0'} + \frac{k+1}{p}. \tag{4.18}$$

Then the k-linear pseudodifferential operator L_σ *can be extended to a bounded operator from* $\mathcal{M}^{p_1,p_1'} \times \mathcal{M}^{p_2,p_2'} \times \cdots \times \mathcal{M}^{p_k,p_k'}$ *into* $\mathcal{M}^{p_0,q_0}$.

We observe again that (4.18) is imposing an obvious upper bound on the integer $k \geq 1$, specifically $k \leq p-1$; thus, also $p \geq 2$. In particular, if $p = 2$, Corollary 4.29 is meaningful only for linear pseudodifferential operators ($k = 1$); or, if we let $p = 3$, then the pseudodifferential operator is either linear or bilinear. For the sake of completeness, let us write down the specifics of the case where $p = 3$ and $k = 2$. Then, $p_0 = 1$ and $q_0 = 1$, and our corollary would read as follows.

Corollary 4.30 *Let $\sigma \in \mathcal{M}^{3,1}(\mathbb{R}^{3d})$ and $1 \leq p_j \leq \infty$, $1 \leq j \leq 2$, be such that*

$$\frac{1}{p_1} + \frac{1}{p_2} = \frac{2}{3}.$$

Then the bilinear pseudodifferential operator L_σ can be extended to a bounded operator from $\mathcal{M}^{p_1,p_1'} \times \mathcal{M}^{p_2,p_2'}$ into $\mathcal{M}^{1,1}$. In particular, L_σ is bounded from $\mathcal{M}^{3,3/2} \times \mathcal{M}^{3,3/2}$ into $\mathcal{M}^{1,1}$.

Finally, let us comment on the statement of Theorem 4.19 when $k = 1$ (linear pseudodifferential operators). Our result now partially recovers a result due to Toft [220, Theorem 4.3].

Corollary 4.31 *Let $\sigma \in \mathcal{M}^{p,1}(\mathbb{R}^{2d})$ and $1 \leq p, p_j, q_j \leq \infty$, $0 \leq j \leq 1$, be such that*

$$\frac{1}{p_0} - \frac{1}{p_1} = \frac{1}{q_0} - \frac{1}{q_1} = \frac{1}{p}.$$

Then the linear pseudodifferential operator L_σ can be extended to a bounded operator from $\mathcal{M}^{p_1,q_1}$ into $\mathcal{M}^{p_0,q_0}$.

Corollary 4.31 is by no means optimal. Indeed, a result of Cordero et al. [81, Theorem 1.1] provides the complete characterization of linear pseudodifferential operators acting between modulation spaces. This, in turn, follows from some optimal continuity results for Fourier integral operators with symbols belonging to appropriate modulation classes, with phase functions that are tame, and acting between modulation spaces. For more on the exciting developments around the topic of Fourier integral operators on modulation spaces, see for example [59, 76–78, 80]. We conclude this section with the optimal boundedness result for linear pseudodifferential operators; see again [81, Theorem 1.1].

Theorem 4.32 *Let $\sigma \in \mathcal{M}^{p,q}(\mathbb{R}^{2d})$ and $1 \leq p, q, p_j, q_j \leq \infty$, $0 \leq j \leq 1$. Then the linear pseudodifferential operator L_σ can be extended to a bounded operator from $\mathcal{M}^{p_1,q_1}$ into $\mathcal{M}^{p_0,q_0}$ if and only if*

$$\max\left(\frac{1}{p_0} - \frac{1}{p_1}, \frac{1}{q_0} - \frac{1}{q_1}\right) \leq \frac{1}{p} - \frac{1}{q'} \quad \text{and} \quad q \leq \min(p_0, q_0, p_1', q_1'). \tag{4.19}$$

4.5 Optimal Boundedness on L^2: The Linear Case

The statement of Theorem 4.32 for $p_1 = q_1 = 2$ and $p_0 = q_0 = 2$ is due to Gröchenig and Heil [129, Theorems 5 and 6], who proved the following particular sharp boundedness result.

Corollary 4.33 *Let $\sigma \in \mathscr{M}^{p,q}(\mathbb{R}^{2d})$, $1 \le p, q \le \infty$. Then the linear pseudodifferential operator L_σ can be extended to a bounded operator on $L^2(\mathbb{R}^d)$ if and only if $p \le q'$ and $q \le 2$.*

A picture of the "admissible" region of indices (p, q) is given in Fig. 4.1.

This section is devoted to a proof of the "only if" part of Corollary 4.33. Specifically, we will prove the following statement, see [129, Theorem 6].

Proposition 4.34

(a) *If $q > 2$, then for any $p \in [1, \infty]$ there exists $\sigma \in \mathscr{M}^{p,q}(\mathbb{R}^{2d})$ such that L_σ is unbounded on $L^2(\mathbb{R}^d)$.*

(b) *If $p > q' \ge 2$, then there exists $\sigma \in \mathscr{M}^{p,q}(\mathbb{R}^{2d})$ such that L_σ is unbounded on $L^2(\mathbb{R}^d)$.*

Our argument will follow closely the one from [129]. However, unlike the proof in [129], we will use throughout the Kohn-Nirenberg correspondence (4.1) that defines the pseudodifferential operator L_σ. In particular, in the proof of Proposition 4.34 (a) we prefer to use the cross-Rychaczek distribution W_1 defined by (4.2)

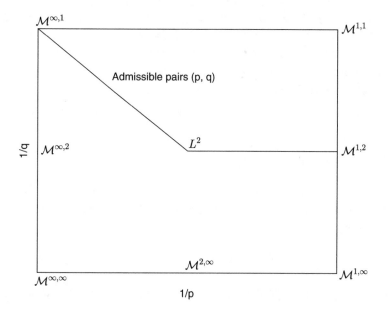

Fig. 4.1 Admissible pairs (p, q)

rather than the Wigner distribution (and consequently the Weyl correspondence) used by the authors in [129]; see Sect. 4.2.

Recall that by (4.2), we have

$$\langle L_\sigma(f), g \rangle = \langle \sigma, W_1(g, f) \rangle,$$

where, for $f, g \in \mathscr{S}$, $W_1(g, f)(x, \xi) = g(x)\overline{\widehat{f}(\xi)}e^{-2\pi i x \cdot \xi}$ is the cross-Rychaczek distribution of f and g.

Let now $\sigma = W_1(\varphi, \psi)$, for some appropriate functions φ and ψ to be chosen later. Then, using Parsevals's identity (Proposition 1.51 (a)), we get

$$\langle L_\sigma(f), g \rangle = \langle W_1(\varphi, \psi), W_1(g, f) \rangle = \langle \varphi, g \rangle \langle \widehat{f}, \widehat{\psi} \rangle = \langle \langle f, \psi \rangle \varphi, g \rangle.$$

In other words, we have proved the following simple fact.

Lemma 4.35 *If $\sigma = W_1(\varphi, \psi)$, then L_σ is a rank-one operator:*

$$L_\sigma(f) = \langle f, \psi \rangle \varphi.$$

The previous lemma is very convenient for our purposes. Since $q > 2$, we know that $L^2(\mathbb{R}^d)$ is strictly contained in $\mathscr{M}^{q,q}(\mathbb{R}^d)$. Select then $\psi \in \mathscr{S}$ such that $\psi \in \mathscr{M}^{q,q}(\mathbb{R}^d) \setminus L^2(\mathbb{R}^d)$, and let $\varphi \in \mathscr{S}$ be arbitrary; in particular, $\varphi \in \mathscr{M}^{p,p}$ for all $p \in [1, \infty]$. Let also $\sigma = W_1(\varphi, \psi)$. By Lemma 4.35, we clearly have L_σ unbounded on $L^2(\mathbb{R}^d)$. Thus, in our search for a counterexample, we would like to create a symbol that necessarily belongs to $\mathscr{M}^{p,q}$ through the cross-Rychaczek distribution $W_1(\varphi, \psi)$ with $\varphi \in \mathscr{M}^{p,p}$ and $\psi \in \mathscr{M}^{q,q}$. We prove that we can indeed do so in the next lemma, which has a strong resemblance of [129, Lemma 7]; this finishes the proof of Proposition 4.34 (a) for $p \in [1, q]$.

Lemma 4.36 *Let $1 \le p \le q \le \infty$ be given. If $\varphi \in \mathscr{M}^{p,p}(\mathbb{R}^d)$ and $\psi \in \mathscr{M}^{q,q}(\mathbb{R}^d)$, then $W_1(\varphi, \psi) \in \mathscr{M}^{p,q}(\mathbb{R}^d)$.*

Proof Fix the nonzero window function $\phi \in \mathscr{S}(\mathbb{R}^d)$. By Lemma 4.17, with $k = 1$, we have

$$\left| V_{W_1(\phi, \phi)} W_1(\varphi, \psi)((u_1, u_2), (v_1, v_2)) \right| = \left| V_\phi \varphi(u_1, v_1 + u_2) \right| \left| V_\phi \psi(u_1 + v_2, u_2) \right|.$$

This nice splitting allows us to estimate rather easily the $\mathscr{M}^{p,q}$ norm of $W_1(\varphi, \psi)$. Let $q < \infty$. Recall also the notation $\tilde{h}(z) = h(-z)$. We have the following sequence of equalities:

$$\|W_1(\varphi, \psi)\|_{\mathscr{M}^{p,q}(\mathbb{R}^d)} = \|V_{W_1(\phi, \phi)} W_1(\varphi, \psi)\|_{L^{p,q}(\mathbb{R}^{2d})}$$

$$= \left(\int_{\mathbb{R}^{2d}} \left(\int_{\mathbb{R}^{2d}} \left| V_{W_1(\phi, \phi)} W_1(\varphi, \psi)((u_1, u_2), (v_1, v_2)) \right|^p du_1 du_2 \right)^{q/p} dv_1 dv_2 \right)^{1/q}$$

$$= \left(\int_{\mathbb{R}^{2d}} \left(\int_{\mathbb{R}^{2d}} |V_\phi \varphi(u_1, v_1 + u_2)|^p |V_\phi \psi(u_1 + v_2, u_2)|^p \, du_1 du_2 \right)^{q/p} dv_1 dv_2 \right)^{1/q}$$

$$= \left(\int_{\mathbb{R}^{2d}} \left(\int_{\mathbb{R}^{2d}} |V_\phi \varphi(u_1, u_2)|^p |V_\phi \psi(u_1 + v_2, u_2 - v_1)|^p \, du_1 du_2 \right)^{q/p} dv_1 dv_2 \right)^{1/q}$$

$$= \left(\int_{\mathbb{R}^{2d}} \left(|V_\phi \varphi|^p * |\widetilde{V_\phi \psi}|^p (-v_2, v_1) \right)^{q/p} dv_1 dv_2 \right)^{1/q}$$

$$= \left\| |V_\phi \varphi|^p * |\widetilde{V_\phi \psi}|^p \right\|_{L^{q/p}}^{1/p}.$$

If we now recall that, by assumption, $\varphi \in \mathcal{M}^{p,p}$, $\psi \in \mathcal{M}^{q,q}$, and $q/p \geq 1$, using Young's inequality (see Theorem 1.11), we can further write

$$\|W_1(\varphi, \psi)\|_{\mathcal{M}^{p,q}(\mathbb{R}^d)} = \left\| |V_\phi \varphi|^p * |\widetilde{V_\phi \psi}|^p \right\|_{L^{q/p}}^{1/p}$$

$$\leq \left\| |V_\phi \varphi|^p \right\|_{L^1}^{1/p} \left\| |\widetilde{V_\phi \psi}|^p \right\|_{L^{q/p}}^{1/p}$$

$$= \|V_\phi \varphi\|_{L^p} \|V_\phi \psi\|_{L^q}$$

$$= \|\varphi\|_{\mathcal{M}^{p,p}} \|\psi\|_{\mathcal{M}^{q,q}} < \infty.$$

The case $q = \infty$ follows from similar calculations. □

Before we begin our proof of part (b), let us point out that the proof of Proposition 4.34 (a) for $p > q$ is also covered by the argument given for part (b). Note also that the assumption on the indices p, q in part (b) is now $p > q' \geq 2$. In particular, we have $p > 2$ and $p' < q \leq 2$.

The counterexample symbol that we are trying to construct will have the particular form $\sigma = \sigma_1 \otimes \sigma_2$. Recall that, by Example 4.6, if σ has this special form, then the linear pseudodifferential operator L_σ is a product-convolution operator, that is,

$$L_\sigma(f) = \sigma_1(\sigma_2^\vee * f) = \sigma_1(f * \mathscr{F}^{-1}\sigma_2).$$

Following [129], we will further look for a tensor like symbol $\sigma = \sigma_1 \otimes \sigma_2$ in which σ_1 and $\sigma_2^\vee$ are defined through appropriate Gabor sums as in (4.20).

We begin with two preliminary lemmas; see [129, Lemma 10].

Lemma 4.37 *Let ϕ be with compact support. Given the sequences* $\mathbf{a} = (a_n)_{n \in \mathbb{Z}^d}$, $\mathbf{b} = (b_n)_{n \in \mathbb{Z}^d}$, *and* $\mathbf{c} = (c_n)_{n \in \mathbb{Z}^d}$, *let*

$$\sigma_1 = \sum_{k \in \mathbb{Z}^d} a_k T_k \phi, \quad \sigma_2 = \sum_{k \in \mathbb{Z}^d} b_k M_{-k} \hat{\phi}, \quad and \quad f = \sum_{k \in \mathbb{Z}^d} c_k T_k \phi. \quad (4.20)$$

If $\sigma = \sigma_1 \otimes \sigma_2$, then

$$L_\sigma(f) = \sum_{|k|\leq K} \sum_{l\in\mathbb{Z}^d} a_{l+k}(\mathbf{b}*\mathbf{c})_l T_l(\phi*\phi) T_{l+k}\phi, \tag{4.21}$$

where K is some positive constant depending only on the size of the support of ϕ.

Proof Observe first that

$$\begin{aligned}
\sigma_2^\vee * f &= \sum_{n\in\mathbb{Z}^d} \sum_{k\in\mathbb{Z}^d} c_n b_k (T_n\phi * T_k\phi) \\
&= \sum_{l\in\mathbb{Z}^d} \Big(\sum_{n\in\mathbb{Z}^d} c_n b_{l-n} \Big) T_l(\phi*\phi) \\
&= \sum_{l\in\mathbb{Z}^d} (\mathbf{b}*\mathbf{c})_l T_l(\phi*\phi).
\end{aligned}$$

Since ϕ has compact support, we can find a $K > 0$ (depending only on the size of the support of ϕ) such that $T_l(\phi*\phi)T_k\phi = 0$ whenever $|l-k| > K$. Therefore,

$$\begin{aligned}
L_\sigma(f) &= \sigma_1(\sigma_2^\vee * f) \\
&= \sum_{k\in\mathbb{Z}^d} \sum_{l\in\mathbb{Z}^d} a_k (\mathbf{b}*\mathbf{c})_l T_l(\phi*\phi) T_k\phi \\
&= \sum_{|k|\leq K} \sum_{l\in\mathbb{Z}^d} a_{l+k}(\mathbf{b}*\mathbf{c})_l T_l(\phi*\phi) T_{l+k}\phi.
\end{aligned}$$

$\square$

Lemma 4.38 *Let $p \in (2,\infty]$ and $q \in (p',2)$. Let $\phi \in \mathcal{M}^{1,1}$ be with compact support. For $\mathbf{a} \in \ell^p(\mathbb{Z}^d)$, $\mathbf{b} \in \ell^q(\mathbb{Z}^d)$, $\mathbf{c} \in \ell^2(\mathbb{Z}^d)$, let $\sigma_1 \in \mathcal{M}^{p,q}$, $\sigma_2 \in \mathcal{M}^{p,q}$, and $f \in L^2$ be as in (4.20). If $\sigma = \sigma_1 \otimes \sigma_2$, then $L_\sigma(f) \in \mathcal{M}^{r,s}$ for some $r \in (2,\infty)$ and all $s \in [1,\infty]$; the convergence of the series (4.21) representing $L_\sigma(f)$ is weak* for $s = \infty$.*

Proof In what follows, we assume $s < \infty$; when $s = \infty$ the word "convergence" should be replaced by "weak* convergence."

By Young's inequality (1.7), we know that $\mathbf{b}*\mathbf{c} \in \ell^q * \ell^2 \subset \ell^t$ where $t \in (2,\infty)$ is defined by $1+\frac{1}{t} = \frac{1}{q}+\frac{1}{2}$, that is, $t = \frac{2q}{2-q}$. Since $\phi \in \mathcal{M}^{1,1}$, by Proposition 2.24 we also have $\phi*\phi \in \mathcal{M}^{1,1}$. Using now Proposition 2.29, we obtain the convergence in $\mathcal{M}^{t,s}$ of $\sum_l (\mathbf{b}*\mathbf{c})_l T_l(\phi*\phi)$. Let now $j \in \mathbb{Z}^d$ be fixed, and define the translation by j sequence $T_j\mathbf{a} = (a_{j+k})_{k\in\mathbb{Z}^d}$. Clearly, $\mathbf{a} \in \ell^p$ implies $T_j\mathbf{a} \in \ell^p$. Therefore, by Hölder's inequality (1.1), we get $(T_j\mathbf{a})(\mathbf{b}*\mathbf{c}) \in \ell^r$, where $r \in (2,\infty)$ is defined by $1/r = 1/p + 1/t = 1/p + 1/q - 1/2$. Using Proposition 2.24 we also have $(T_j\phi)(\phi*\phi) \in \mathcal{M}^{1,1}$; thus another application of Proposition 2.29 will now give the convergence in $\mathcal{M}^{r,s}$ of $\sum_l \big((T_j\mathbf{a})(\mathbf{b}*\mathbf{c})\big)_l T_l\big((T_j\phi)(\phi*\phi)\big)$, which is the same as $\sum_{l\in\mathbb{Z}^d} a_{l+j}(\mathbf{b}*\mathbf{c})_l T_l(\phi*\phi)T_{l+j}\phi$. Finally, using (4.21) in Lemma 4.37, we conclude that $L_\sigma \in \mathcal{M}^{r,s}$ for some $r \in (2,\infty)$ and all $s \in [1,\infty]$. $\square$

We are now finally ready for the construction of our counterexample. In the spirit of our previous two lemmas, we will select $\mathbb{Q} = [-\frac{1}{2}, \frac{1}{2}]^d$ and let

$$\phi = \chi_{\mathbb{Q}} * \chi_{\mathbb{Q}}.$$

We note immediately that ϕ is compactly supported and $\hat{\phi} \in L^1(\mathbb{R}^d)$. Therefore, by Proposition 2.24, we see that $\phi \in \mathscr{M}^{1,1}$. Moreover, $\phi \geq 0$ and, for some constant $c > 0$, we have

$$\phi(\phi * \phi) \gtrsim \chi_{(2c)^d \mathbb{Q}}. \tag{4.22}$$

Let also $\mathbf{a} \in \ell^p$, $\mathbf{b} \in \ell^q$, $\mathbf{c} \in \ell^2$ be three sequences of positive numbers, and σ and f be defined as in Lemma 4.37. By Lemma 2.10, we have $\sigma \in \mathscr{M}^{p,q}$, while by Lemma 4.38 we have $L_\sigma(f) \in \mathscr{M}^{r,r}$ for some $r \in (2, \infty)$. In particular, since we have the strict inclusion $L^2 = \mathscr{M}^{2,2} \subset \mathscr{M}^{r,r}$ for $r > 2$, we would not be wrong to claim that $L_\sigma(f)$ does not necessarily belong to L^2 even though f does.

The last part of our argument specifically constructs some positive sequences $\mathbf{a} \in \ell^p$, $\mathbf{b} \in \ell^q$, $\mathbf{c} \in \ell^2$ for which, indeed, $L_\sigma(f) \notin L^2$. Let us observe first that, under the previous assumptions, all terms in the series (4.21) that defines $L_\sigma(f)$ are positive. Then, using (4.22), we get

$$\|L_\sigma(f)\|_{L^2} \geq \left\| \sum_l a_l (\mathbf{b} * \mathbf{c})_l T_l(\phi(\phi * \phi)) \right\|_{L^2} \gtrsim \|\mathbf{a}(\mathbf{b} * \mathbf{c})\|_{\ell^2}.$$

Therefore, the proof of our claim reduces to the construction of three sequences as above that satisfy the additional condition $\mathbf{a}(\mathbf{b} * \mathbf{c}) \notin \ell^2$.

Recall that $q > p'$. Then, $1/p + 1/q + 1/2 < 1/p + 1/p' + 1/2 = 3/2$. Thus, there exists some $\epsilon > 0$ such that $1/p + 1/q + 1/2 < (3d)/(2d + 2\epsilon)$. For $n \in \mathbb{Z}^d \setminus \{0\}$, the n-th terms of our three sequences are given by

$$a_n = |n|^{-\frac{d+\epsilon}{p}}, \quad b_n = |n|^{-\frac{d+\epsilon}{q}} \quad \text{and} \quad c_n = |n|^{-\frac{d+\epsilon}{2}};$$

the zero index terms are chosen to be some arbitrarily positive numbers. Clearly, $\mathbf{a} \in \ell^p(\mathbb{Z}^d)$, $\mathbf{b} \in \ell^q(\mathbb{Z}^d)$, and $\mathbf{c} \in \ell^2(\mathbb{Z}^d)$ are all positive sequences. Fix now $n \in \mathbb{Z}^d \setminus \{0\}$. We have the following straightforward estimate on the n-th term of $\mathbf{b} * \mathbf{c}$:

$$(\mathbf{b} * \mathbf{c})_n \geq \sum_{|k| \sim |n|/4} b_k c_{n-k} \gtrsim |n|^d |n|^{-(d+\epsilon)\left(\frac{1}{p}+\frac{1}{q}\right)}.$$

Thus

$$a_n (\mathbf{b} * \mathbf{c})_n \gtrsim |n|^d |n|^{-(d+\epsilon)\left(\frac{1}{p}+\frac{1}{q}+\frac{1}{2}\right)} \gtrsim |n|^{-|d|/2},$$

which clearly implies that $\mathbf{a}(\mathbf{b} * \mathbf{c}) \notin \ell^2$.

4.6   Abstract Multiplier Theorems via Wiener Amalgams

The general idea throughout the previous sections is that the boundedness of Fourier multipliers or pseudodifferential operators on modulation spaces is related to the localization of the multiplier or symbol in time and frequency. Indeed, there are lots of concrete examples of multipliers that are not bounded on Lebesgues spaces, but bounded on modulation spaces. For example, as explained in [20], a Fourier multiplier L_σ with

$$\sigma = \sum_{n \in \mathbb{Z}^d} c_n \chi_{n + \mathbb{Q}_b},$$

where $(c_n) \in \ell^\infty(\mathbb{Z}^d)$, $b_j > 0$, and $\mathbb{Q}_b = \prod_{j=1}^d (0, b_j)$ will be bounded on all $\mathscr{M}^{p,q}$ spaces with $1 < p < \infty$, $1 \leq q \leq \infty$, but not on L^p, $p \neq 2$, except in trivial cases.

The characterization of Fourier multipliers on Lebesgue spaces is a question of interest in harmonic analysis. We simply recall here that the set of Fourier multipliers on L^p, $1 \leq p \leq \infty$, coincides with that of Fourier multipliers on $L^{p'}$, that Fourier multipliers on L^2 are precisely the ones corresponding to symbols in L^∞, and that the Fourier multipliers on L^1 (hence, also on L^∞) are exactly the ones whose Fourier transform of the multiplier is a (complex-valued) finite Borel measure; see, for example, [119, 120]. Tightly connected with the understanding of all Fourier multipliers on Lebesgue spaces is that of Fourier multipliers on modulation spaces. Indeed, such abstract results first appeared in Feichtinger's seminal paper [94]. Later on, Feichtinger and Narimani [102] provided a "simpler" abstract characterization of Fourier multipliers on $\mathscr{M}^{p,q}(\mathbb{R}^d)$ which relies on a Wiener amalgam space.

Definition 4.39   Let $\phi \in \mathscr{S}(\mathbb{R}^d) \setminus \{0\}$, be a fixed window. The space $W(M_{\mathscr{F}}(L^p), L^\infty)$ consists of all tempered distributions σ for which

$$\|\sigma\|_{W(M_{\mathscr{F}}(L^p), L^\infty)} = \sup_{x \in \mathbb{R}^d} \|L_\sigma T_x \phi\|_{L^p \to L^p} < \infty. \tag{4.23}$$

From Remark 4.8, we recall that, for a Fourier multiplier $\sigma(x, \xi) = \sigma(\xi)$, we have defined $L_\sigma(f) = \sigma^\vee * f$.

With this definition, the main result in [102] is the following.

Theorem 4.40   *Let $1 \leq p, q \leq \infty$. A Fourier multiplier L_σ is bounded on $\mathscr{M}^{p,q}$ if and only if $\sigma \in W(M_{\mathscr{F}}(L^p), L^\infty)$.*

While this result is of interest in its own right, it is difficult to apply it to concrete examples since each $\sigma \in W(M_{\mathscr{F}}(L^p), L^\infty)$ coincides locally with a Fourier multiplier on L^p.

We present now a more "concrete" result [20, 66] which generalizes the linear case of Corollary 4.21 and is useful when dealing with multipliers commonly appearing in PDEs.

Proposition 4.41 *Let* $0 < p, q \leq \infty$. *If* $1 \leq p \leq \infty$ *and* $\sigma \in W(\mathscr{F}L^1, L^\infty)$, *then the Fourier multiplier* L_σ *is bounded on* $\mathscr{M}^{p,q}$. *If* $0 < p < 1$ *and* $\sigma \in W(\mathscr{F}L^p, L^\infty)$, *then* L_σ *is bounded on* $\mathscr{M}^{p,q}$.

We note immediately that since $\mathscr{F}L^1 \subset M_{\mathscr{F}}(L^p)$, Proposition 4.41 is essentially a consequence of Theorem 4.40. However, we present here a simple and direct proof of the case $1 \leq p \leq \infty$. The calculations proving the case $0 < p < 1$ can be found in [14].

Proof Let us write $\phi^*(x) = \bar{\bar{\phi}}(x) = \overline{\phi(-x)}$. Choose then $\phi_0 \in \mathscr{S}(\mathbb{R}^d)$ a nonzero window and let $\phi = \phi_0 * \phi_0$. Using Young's inequality and the fact that $M_\xi \phi = M_\xi \phi_0 * M_\xi \phi_0$, we have the following:

$$\|L_\sigma(f)\|_{\mathscr{M}^{p,q}} = \left(\int_{\mathbb{R}^d} \|(L_\sigma(f)) * (M_\xi \phi^*)\|^q_{L^p} \, d\xi \right)^{1/q}$$

$$= \left(\int_{\mathbb{R}^d} \|(\sigma^\vee * M_\xi \phi_0^*) * (f * M_\xi \phi_0^*)\|^q_{L^p} \, d\xi \right)^{1/q}$$

$$\leq \left(\int_{\mathbb{R}^d} \|\sigma^\vee * M_\xi \phi_0^*\|^q_{L^1} \|f * M_\xi \phi_0^*\|^q_{L^p} \, d\xi \right)^{1/q}$$

$$\leq \left(\sup_{\xi \in \mathbb{R}^d} \|\sigma^\vee * M_\xi \phi_0^*\|_{L^1} \right) \left(\int_{\mathbb{R}^d} \|f * M_\xi \phi_0^*\|^q_{L^p} \, d\xi \right)^{1/q}$$

$$= \|\sigma\|_{W(\mathscr{F}L^1, L^\infty)} \|f\|_{\mathscr{M}^{p,q}}.$$

$\square$

Note also that this Proposition recovers the case $k = 1$ in Corollary 4.21, since clearly $\mathscr{M}^{\infty,1} \subset W(\mathscr{F}L^\infty, L^1)$.

4.7 Schrödinger Multipliers on Modulation Spaces

We begin by recalling that in Sect. 2.3, Example 2.20, we defined the multiplier $\sigma_{t;2}(\xi) = e^{i\pi t |\xi|^2}$, $\xi \in \mathbb{R}^d$ and $t \in \mathbb{R}$, and showed that $\sigma_{t;2} \in (W(\mathscr{F}L^p, L^\infty) \cup \mathscr{M}^{1,\infty}) \setminus \mathscr{M}^{\infty,1}$. Thus, while Corollary 4.21 is not applicable, we can use Proposition 4.41 and we get that the corresponding Fourier multiplier $L_{\sigma_{t;2}}$ is bounded on all $\mathscr{M}^{p,q}$, $0 < p, q \leq \infty$. In fact, this approach works for all *generalized Gaussians* of the form

$$\sigma_{A;b}(\xi) = e^{-\pi \xi \cdot A\xi + 2\pi b \cdot \xi},$$

where $b \in \mathbb{C}^d$ and $A = B + iC$, with B a positive-definite real-valued $d \times d$-matrix and C a symmetric real-valued matrix. Indeed, in this case, we simply need to note that, after a change of coordinates, any quadratic function can be written in the form $\langle \xi, M\xi \rangle$, where M is a $d \times d$ Hermitian matrix; the calculations from Example 2.20 go through almost identically.

The multiplier $\sigma_{t;2}$ arises naturally in the study of the Schrödinger equation, the subject of a significant part of Chap. 7. Naturally, the remarks above lead to the question of whether a similar phenomenon works for other multipliers of Schrödinger type, specifically of the form

$$\sigma_{t;\alpha}(\xi) = e^{i\pi t |\xi|^{\alpha}}, \xi \in \mathbb{R}^d, t \in \mathbb{R}, \alpha \geq 0.$$

The following theorem gives a hint of what might happen when $\alpha \neq 2$; see [20, Corollary 15].

Theorem 4.42 *If $\alpha \in [0, 1]$, then $\sigma_{t;\alpha} = \sigma_{t;(0)} + \sigma_{t;(\infty)}$, where $\sigma_{t;(0)} \in \mathscr{F}L^1(\mathbb{R}^d)$ and $\sigma_{t;(\infty)} \in S^0_{0,0}(\mathbb{R}^d)$. In particular, we have $\sigma_{t;\alpha} \in \mathscr{M}^{\infty,1}(\mathbb{R}^d)$ and $L_{\sigma_{t;\alpha}}$ is bounded on all $\mathscr{M}^{p,q}, 0 < p, q \leq \infty$.*

Proof The parameter t does not play much of a role here, so we let $t = 1$ and we write σ_α for $\sigma_{1;\alpha}$. Let $\chi \in \mathscr{C}^\infty_c(\mathbb{R}^d)$ be a *bump function* such that $\chi(\xi) = 1$ for $|\xi| \leq 1$, $\chi(\xi) = 0$ for $|\xi| \geq 2$, and $0 \leq \chi(\xi) \leq 1$ for $1 \leq |\xi| \leq 2$. Next, we split σ_α into two parts that capture the singularity at the origin and the oscillatory behavior at infinity, respectively:

$$\sigma_\alpha = \chi\sigma_\alpha + (1 - \chi)\sigma_\alpha = \sigma_{(0)} + \sigma_{(\infty)}.$$

Let us first consider the "global part" $\sigma_{(\infty)}$ of the symbol. For $|\xi| \geq 1$ and $|\beta| \geq 1$, straightforward calculations show that

$$|\partial^\beta \sigma_{(\infty)}(\xi)| \lesssim (1 + |\xi|)^{\alpha - |\beta|}.$$

Since $\alpha \in [0, 1]$, we obtain from here that, for all $\xi \in \mathbb{R}^d$, and all multi-indices β, we have

$$|\partial^\beta \sigma_{(\infty)}(\xi)| \lesssim 1.$$

In particular, $\sigma_{(\infty)} \in S^0_{0,0} \subset \mathscr{M}^{\infty,1}$; see again [127, Therem 1.1] or [125, Theorem 14.5.3] for this inclusion.

Next, we consider the "local part" $\sigma_{(0)}$ of the symbol, and show that $\sigma_{(0)} \in \mathscr{F}L^1$. Assuming this is true, and recalling that $\mathscr{F}L^1 \subset \mathscr{M}^{\infty,1}$, see, for example, [20, Lemma 8], the theorem is proved.

We begin by writing

$$\sigma_{(0)}(\xi) = e^{i|\xi|^\alpha} \chi(\xi) = \sum_{k=0}^{\infty} \frac{i^k}{k!} |\xi|^{\alpha k} \chi(\xi).$$

For $k \in \mathbb{N} \cup \{0\}$, let $\phi_k(\xi) = |\xi|^{\alpha k} \chi(\xi)$. Next, for $\xi \neq 0$, we write $\chi(\xi) = \sum_{j=1}^{\infty} \psi(2^j \xi)$, where ψ is a $\mathscr{C}_c^{\infty}(\mathbb{R}^d)$ function supported in the annulus $\{\xi \in \mathbb{R}^d : 1 \leq |\xi| \leq 4\}$. From here, we obtain

$$\phi_k(\xi) = \sum_{j=1}^{\infty} |\xi|^{\alpha k} \psi(2^j \xi)$$

$$= \sum_{j=1}^{\infty} 2^{-kj\alpha} |2^j \xi|^{\alpha k} \psi(2^j \xi)$$

$$= \sum_{j=1}^{\infty} 2^{-kj\alpha} \psi_k(2^j \xi),$$

with $\psi_k(\xi) = |\xi|^{\alpha k} \psi \in \mathscr{C}_c^{\infty}$. Clearly, the $\mathscr{F}L^1$ norm of $\psi_k(2^j \cdot)$ is independent of j, since

$$\|\psi_k(2^j \cdot)\|_{\mathscr{F}L^1} = \|2^{-jd} \widehat{\psi_k}(2^{-j} \cdot)\|_{L^1} = \|\widehat{\psi_k}\|_{L^1} = \|\psi_k\|_{\mathscr{F}L^1}.$$

We will now find estimates on $\|\psi_k\|_{\mathscr{F}L^1}$ that guarantee the convergence in $\mathscr{F}L^1$ of the series $\sum_{k=0}^{\infty} (i^k / k!) \phi_k$ defining $\sigma_{(0)}$. Clearly,

$$\|\psi_k\|_{\mathscr{F}L^1} = \int_{|\xi| \leq 1} |\widehat{\psi_k}(\xi)| \, d\xi + \int_{|\xi| \geq 1} |\widehat{\psi_k}(\xi)| \, d\xi = I + II.$$

To estimate the first term, I, observe that $|\xi|^{k\alpha} \lesssim 4^{\alpha k}$ on the support of ψ. Thus,

$$I \leq |\mathbb{B}(0,1)| \|\widehat{\psi_k}\|_{L^\infty} \lesssim \sup_{|\xi| \leq 4} |\xi|^{\alpha k} \|\psi\|_{L^1} \lesssim 4^{\alpha k}. \tag{4.24}$$

Let us now estimate the second term, II. We note that the following pointwise estimate for indices $|\beta| \leq d + 1$ holds:

$$|\widehat{\psi_k}(\xi)| = |(2\pi i \xi)^{-\beta}| |\widehat{\partial^\beta \psi_k}(\xi)| \lesssim \max_{|\beta| \leq d+1} \|\widehat{\partial^\beta \psi_k}\|_{L^\infty} \min_{|\beta| \leq d+1} |\xi^{-\beta}|.$$

Now, by Leibniz' rule, $\partial^\beta \psi_k(\xi)$ can be expressed as a (finite) sum of terms of the form

$$(\partial^{\gamma_{k+1}}\psi(\xi))\prod_{j=1}^{k}\partial^{\gamma_j}|\xi|^{\alpha k},$$

with $\sum_{j=1}^{k+1}\gamma_j = \beta$. Clearly, we have

$$|\partial^{\gamma_j}|\xi|^{\alpha k}| \lesssim 4^{\alpha-|\gamma_j|} \le 4^{\alpha k} \text{ and } |\partial^{\gamma_{k+1}}\psi(\xi)| \lesssim 1.$$

Putting all these estimates together, we arrive at

$$II \lesssim \max_{|\beta|\le d+1}\|\partial^\beta\psi_k\|_{L^1}\int_{|\xi|\ge 1}\min_{|\beta|\le d+1}\frac{1}{|\xi^\beta|}\,d\xi.$$

Since $\min_{|\beta|\le d+1}\dfrac{1}{|\xi^\beta|}$ is integrable outside the unit ball, see [125, p. 321],

$$II \lesssim 4^{\alpha k}. \tag{4.25}$$

Therefore, from (4.24) and (4.25), we conclude that

$$\|\psi_k\|_{\mathscr{F}L^1} \lesssim 4^{\alpha k},$$

and

$$\|\phi_k\|_{\mathscr{F}L^1} \le \sum_{j=1}^{\infty}2^{-kj\alpha}\|\psi_k\|_{\mathscr{F}L^1} \lesssim \frac{2^{\alpha k}}{1-2^{-\alpha k}}.$$

Finally, recalling the series representation of $\sigma_{(0)}$, we get

$$\|\sigma_{(0)}\|_{\mathscr{F}L^1} \le \sum_{k=0}^{\infty}\frac{1}{k!}\|\phi_k\|_{\mathscr{F}L^1} \lesssim \sum_{k=0}^{\infty}\frac{2^{\alpha k}}{(1-2^{-\alpha k})k!} < \infty.$$

□

A natural question is whether Theorem 4.42 works for values of α that are greater than 1. We saw that this is indeed the case for $\alpha = 2$. Although there are some changes to be made in the case where $\alpha > 1$, the idea of splitting the symbols σ_α into their local and global components from the proof of Theorem 4.42 can be used again. For example, if $\alpha \in (1, 2]$, the main observation is that now $\sigma_{(\infty)} \in W(\mathscr{F}L^1, L^\infty)$. By Proposition 4.41, we have yet again the boundedness of L_{σ_α} on all $\mathscr{M}^{p,q}$ spaces. The situation for $\alpha > 2$ is also well understood. While one still expects boundedness on modulation spaces, this comes at the expense of some loss of derivatives, which in turn can be understood as the boundedness of L_{σ_α} from a weighted modulation space $\mathscr{M}^{p,q}_s$ (see Chap. 5 for the precise definition) into a modulation space $\mathscr{M}^{p,q}$. More precisely, we have the following result due to Miyachi et al. [184, Theorem 1.1].

Theorem 4.43 *Let $\alpha > 2$, $1 \leq p, q \leq \infty$, and $s > (\alpha - 2)d|1/p - 1/2|$. Then $L_{\sigma_{t;\alpha}}$ is bounded from $\mathscr{M}^{p,q}_s(\mathbb{R}^d)$ to $\mathscr{M}^{p,q}(\mathbb{R}^d)$.*

The loss of regularity $s_0 = (\alpha - 2)d|1/p - 1/2|$ is essentially sharp, see [184, Theorem 1.2]. Furthermore, one can prove more exact estimates capturing the t dependence of the operator norm of the multiplier $L_{\sigma_{t;\alpha}}$. The following result is due to Chen et al. [45, Theorem 1 and Theorem 2]; see also [207, Theorem 1.1] for an extension.

Theorem 4.44 *Let s_0 be defined as above, $s > s_0$ and $0 < p, q \leq \infty$.*

(a) *If $\alpha \in (1, 2]$, then $\|L_{\sigma_{t;\alpha}}(f)\|_{\mathscr{M}^{p,q}_s} \lesssim \langle t \rangle^{d|1/2-1/p|} \|f\|_{\mathscr{M}^{p,q}_s}$.*

(b) *If $\alpha > 2$, then $\|L_{\sigma_{t;\alpha}}(f)\|_{\mathscr{M}^{p,q}_{s-s_0}} \lesssim \langle t \rangle^{d|1/2-1/p|} \|f\|_{\mathscr{M}^{p,q}_s}$.*

The results discussed in this section are concerned with particular types of *unimodular Fourier multipliers*. Indeed, the result in Theorem 4.42 and its proof are simply adapted from the more general version that holds for such multipliers; see also [90] and [224] for similar results. For variations of the results pointed out in Theorem 4.44 that hold on modulation-Hardy spaces or α-modulation spaces, see [44, 46, 47] and [48]. For a discussion of the time-frequency analysis of Schrödinger propagators, including further asymptotic estimates for the solution of the nonlinear Schrödinger equation with certain potentials, see [79, 80, 82, 83] and [160–163]. A treatment of the mapping properties between modulation spaces for very general operators of the form e^{itH}, with H being the sum of a pseudodifferential operator with symbol in a suitable modulation space and the Weyl quantization of a real quadratic homogeneous polynomial, can be found in [68].

4.8 Notes

We provide here a brief discussion about localization operators since they tie in nicely with pseudodifferential operators both in the linear and multilinear setting. *Localization operators*, also known as *anti-Wick operators* or *Toeplitz operators*, were introduced by Berezin [25] as a quantization rule in physics and then took a time-frequency life of their own through Daubechies's work [87].

Given a symbol $a \in \mathscr{S}'(\mathbb{R}^{2kd})$ and the window functions $\phi_j \in \mathscr{S}(\mathbb{R}^d)$, $1 \leq j \leq 2k$, the k-linear localization operator A_a is the mapping defined a priori from $\underbrace{\mathscr{S}(\mathbb{R}^d) \times \mathscr{S}(\mathbb{R}^d) \times \cdots \times \mathscr{S}(\mathbb{R}^d)}_{k \text{ times}}$ to $\mathscr{S}'(\mathbb{R}^d)$ via

$$A_a(\overrightarrow{f})(x) = \int_{\mathbb{R}^{2kd}} a(\overrightarrow{y}, \overrightarrow{\xi}) V_{\otimes_{j=1}^k \phi_j}(\otimes_{j=1}^k f_j)(\overrightarrow{y}, \overrightarrow{\xi}) \prod_{j=1}^m M_{\xi_j} T_{y_j} \phi_{m+j}(x) \, d\overrightarrow{\xi} \, d\overrightarrow{y},$$

where, as usual, we denoted $\overrightarrow{f} = (f_1, f_2, \ldots, f_k)$, $\overrightarrow{y} = (y_1, y_2, \ldots, y_k) \in \mathbb{R}^{kd}$ and $\overrightarrow{\xi} = (\xi_1, \xi_2, \ldots, \xi_k) \in \mathbb{R}^{kd}$. The connection with multilinear pseudodifferential operators is the following: we have $A_a = L_\sigma$, where the symbol σ is given by

$$\sigma(x, \overrightarrow{\xi}) = (a * \Phi)(X, \overrightarrow{\xi});$$

here, $X = \underbrace{(x, x, \ldots, x)}_{k \text{ times}}$ and $\Phi(\overrightarrow{y}, \overrightarrow{\xi}) = \prod_{j=1}^k \widehat{V_{\phi_j} \phi_{j+m}}(-\xi_j, y_j)$. As shown by Cordero and Okoudjou [70], this connection and the now known boundedness properties of multilinear pseudodifferential operators on modulation spaces yield boundedness results for these multilinear localization operators; in particular, the convolution of an anti-Wick symbol $a \in \mathscr{M}^{\infty,\infty}(\mathbb{R}^{2kd})$ with a suitable Φ yields Kohn-Nirenberg symbols $\sigma \in \mathscr{M}^{\infty,1}$. Only to give a flavor of what happens for these operators, we mention the following result from [70]: if $a \in \mathscr{M}^{\infty,\infty}(\mathbb{R}^{2kd})$ and $\phi_j \in \mathscr{M}^{1,1}(\mathbb{R}^d)$, then A_a extends to a bounded operator from $\mathscr{M}^{p_1,q_1} \times \mathscr{M}^{p_2,q_2} \times \cdots \times \mathscr{M}^{p_k,q_k}$ to $\mathscr{M}^{p_0,q_0}$, where the exponents p_j, q_j satisfy the conditions from Proposition 4.18. Nowadays we have a well-developed theory of the localization operators in both linear and multilinear setting comprising regularity properties, such as Schatten-von Neumann class and compactness properties, or a symbolic calculus for their composition and the Fredholm property. A nice survey about localization operators in time-frequency analysis is given in [74], while a unification of various localization operators in the context of homogeneous spaces can be found in [231]. For further works on the topic of localization operators, the interested reader can consult, for example, [30–33, 60–63, 67, 71–73, 106, 107, 131].

Chapter 5
Weighted Modulation Spaces

The concept of *weight function* appears ubiquitously in harmonic analysis. A weight is a nonnegative measurable function that, depending on the context, quantifies more precisely growth, decay, or smoothness. A classic example in the linear and multilinear Calderón-Zygmund theory is played by the nowadays well-understood A_p classes of weights of Muckenhoupt-Wheeden which give natural weighted norm inequalities on Lebesgue spaces for the maximal operator, singular integrals, and much more [186–188]. The A_p-weights are intrinsically connected to reverse Hölder inequalities and they were from their inception important in the theory of conformal mappings [117] and boundary-value problems for the Laplace equation on a bounded domain with Lipschitz boundary [85, 86]; see [115] for more details about these topics. The special interest in the A_2-class and the subsequent settling of the so-called A_2-conjecture by Hytönen [156] came amid a flurry of works that introduced new ideas and techniques to this area of research.

In the context of time-frequency analysis as understood in this text, it is rather fascinating to see them in the study of Gabor systems via the Zak transform [143] given that a major drawback of the Muckenhoupt-Wheeden classes of weights has to do with the non-invariance under time-frequency shifts of the corresponding weighted modulation spaces. As it turns out, the role of the correct class for the study of weighted modulation spaces is played by that of *moderate weights*, with respect to some given sub-multiplicative weight, that capture the time-frequency concentrations of the appropriate signals in these spaces. A relevant example in the study of partial differential equations is played by those moderate weights with respect to a submultiplicative weight that grows at most polynomially and it is in this context that we find it useful to study the weighted modulation spaces $\mathcal{M}_s^{p,q}$. Their multiplication algebra property for appropriate parameters p, q, s, in particular, becomes crucial in controlling certain nonlinear terms and leads to beautiful questions about operating functions on modulation spaces that have direct consequences for our discussion of certain dispersive partial differential equations in Chap. 7. Furthermore, the moderate weights considered here capture the smoothing

© Springer Science+Business Media, LLC, part of Springer Nature 2020
Á. Bényi, K. A. Okoudjou, *Modulation Spaces*, Applied and Numerical
Harmonic Analysis, https://doi.org/10.1007/978-1-0716-0332-1_5

properties of certain classes of pseudodifferential operators as explained already at the end of Chap. 4 and they help elucidate the embeddings between modulation spaces and the other classical function spaces of analysis in Chap. 6. A nice survey of the most relevant classes of weights appearing in time-frequency analysis with an emphasis on the roles played in defining a concept or solving a given problem can be found in [126].

5.1 Definition and the Lifting Property

Recall the "Japanese bracket" notation $\langle \xi \rangle = (1 + 4\pi^2 |\xi|^2)^{1/2}$. Let $0 < p, q \le \infty$ and $s \in \mathbb{R}$. Consider the Fourier multiplier operator with symbol $\sigma_s(\xi) = \langle \xi \rangle^s$, that is, for say $f \in \mathscr{S}$,

$$L_{\sigma_s}(f) = f * \sigma_s^\vee \quad \text{or} \quad \widehat{L_{\sigma_s}(f)} = \sigma_s \widehat{f}.$$

The multiplier L_{σ_s} is typically denoted by $(I - \Delta)^{s/2}$. It is easy to see that for all $s, t \in \mathbb{R}$ we have $L_{\sigma_s} L_{\sigma_t} = L_{\sigma_{s+t}}$ and $L_{\sigma_0} = I$; in particular, the inverse of L_{σ_s} is $L_{\sigma_{-s}}$. Thus $\{L_{\sigma_s}\}_{s \in \mathbb{R}}$ is a group of operators (acting on $\mathscr{S}$ or $\mathscr{S}'$).

Definition 5.1 Let $\phi \in \mathscr{S}(\mathbb{R}^d) \setminus \{0\}$. The modulation space $\mathscr{M}_s^{p,q} = \mathscr{M}_s^{p,q}(\mathbb{R}^d)$ is the set of all tempered distribution $f \in \mathscr{S}'(\mathbb{R}^d)$ for which

$$\|f\|_{\mathscr{M}_s^{p,q}} := \left(\int_{\mathbb{R}^d} \left(\int_{\mathbb{R}^d} |V_\phi f(x, \xi) \sigma_s(\xi)|^p dx \right)^{q/p} d\xi \right)^{1/q} < \infty. \tag{5.1}$$

When $p = \infty$ or $q = \infty$, the essential supremum is used.

In other words,

$$\|f\|_{\mathscr{M}_s^{p,q}} = \|V_\phi f(x, \xi)\|_{L^p(dx) L^q(\langle \xi \rangle^s d\xi)}.$$

We have the following *lifting property*; see [94, Theorem 6.1], also [219, Corollary 3.3].

Theorem 5.2 *The mapping* $L_{\sigma_{s-t}}$ *is an isomorphism between* $\mathscr{M}_t^{p,q}$ *and* $\mathscr{M}_s^{p,q}$.

Proof Note that due to the group property of the family $\{L_{\sigma_s}\}_{s \in \mathbb{R}}$, it suffices to show that L_{σ_s} is an isomorphism between $\mathscr{M}^{p,q}$ and $\mathscr{M}_s^{p,q}$. Specifically, we will show that if $f \in \mathscr{M}_s^{p,q}$, then $L_{\sigma_s}(f) \in \mathscr{M}^{p,q}$. Let $\phi \in \mathscr{S}$ with $\|\phi\|_{L^2} = 1$ be fixed and $\Psi(x, y) = \sigma_s(x) \sigma_{-s}(y) \sigma_{-N}(x - y)$ with $N = |s| + d + 1$. Denote by Ψ_2 the partial Fourier transform of $\Psi(x, y)$ in the y-variable. Let also $\phi_0(\xi) = \langle \xi \rangle^N \widehat{\phi}(\xi)$ and $u(y) = \langle y \rangle^N \sup_\xi |\Psi_2(\xi, y)|$. Note that $u \in L^1$ by Toft [219, Lemma 3.1]. Now, by Parseval's identity, we can write

$$|\mathcal{F}(L_{\sigma_s}(fT_x\check{\phi})(\xi)| = |\mathcal{F}((\cdot)^s \widehat{f} T_\xi \widehat{\phi})(-x)|$$
$$= \sigma_s(\xi)|\mathcal{F}(\widehat{f}T_\xi\phi_0\Psi(\xi,\cdot))(-x)|$$
$$= \sigma_s(\xi)\left|\int_{\mathbb{R}^d} \mathcal{F}(\widehat{f}T_\xi\phi_0)(-x-y)\Psi_2(\xi,y)\,dy\right|.$$

By applying now Minkowski's integral inequality-Theorem 1.3, we finally obtain

$$\|L_{\sigma_s}(f)\|_{\mathcal{M}^{p,q}}$$
$$\lesssim \left(\int \left(\int \left(\int |\mathcal{F}(fT_{x+y}\tilde{\phi}_0)(\xi)\Psi_2(\xi,y)\langle\xi\rangle^s|\,dy\right)^p dx\right)^{q/p} d\xi\right)^{1/q}$$
$$\lesssim \|u\|_{L^1}\left(\int \left(\int |\mathcal{F}(fT_x\tilde{\phi}_0)(\xi)\sigma_s(\xi)|^p dx\right)^{q/p} d\xi\right)^{1/q}$$
$$\lesssim \|u\|_{L^1}\|f\|_{\mathcal{M}_s^{p,q}}.$$

$\square$

5.2 Properties of Weighted Modulation Spaces

Proposition 5.3

(a) *The definition of the modulation space $\mathcal{M}_s^{p,q}$ is independent of the window $\phi \in \mathcal{S}(\mathbb{R}^d)$.*

(b) *The modulation spaces $\mathcal{M}_s^{p,q}$ are quasi-Banach spaces.*

(c) *We have that $\mathcal{S} \subset \mathcal{M}_s^{p,q}$. Moreover, if $p, q < \infty$, then $\mathcal{S}$ is dense in $\mathcal{M}_s^{p,q}$.*

(d) *If $0 < p_1 \le p_2$ and $0 < q_1 \le q_2$, then $\mathcal{M}_s^{p_1,q_1} \subseteq \mathcal{M}_s^{p_2,q_2}$.*

(e) *If $0 < q_1 \le q_2$ and $(s_1-s_2)/d > 1/q_2 - 1/q_1$, then $\mathcal{M}_{s_1}^{p,q_1}(\mathbb{R}^d) \subseteq \mathcal{M}_{s_2}^{p,q_2}(\mathbb{R}^d)$. In particular, if $s > d/q'$, then $\mathcal{M}_s^{p,q} \subset \mathcal{M}^{\infty,1}$.*

(f) *We have the following duality relation: $(\mathcal{M}_s^{p,q})' = \mathcal{M}_{-s}^{p',q'}$.*

Proof The properties follow naturally from the lifting property and the already proved basic properties of unweighted modulation spaces $\mathcal{M}^{p,q}$. $\square$

Recall that, as explained in Chap. 2, $\mathcal{M}^{2,2} = L^2$. The definition of $\mathcal{M}_s^{p,q}$ immediately shows that we also have the identification $\mathcal{M}_s^{2,2} = L_s^2 = H^s$. As for the classical potential spaces L_s^p, the weighted modulation spaces $\mathcal{M}_s^{p,q}$ turn out to be multiplication algebras; see Proposition 1.63. More precisely, we have the following statement.

Proposition 5.4 *Let $0 < p, q \le \infty$ and $s > d/q'$. Then, for all $f, g \in \mathcal{M}_s^{p,q}(\mathbb{R}^d)$ we have*

$$\|fg\|_{\mathscr{M}_s^{p,q}} \lesssim \|f\|_{\mathscr{M}_s^{p,q}} \|g\|_{\mathscr{M}_s^{p,q}}.$$

Proof Note first that for $\phi \in \mathscr{S}$ we can write

$$V_{\phi^2}(fg)(x,\xi) = \int_{\mathbb{R}^d} V_\phi f(x, \xi - \eta) V_\phi g(x, \eta)\, d\eta.$$

Thus, as a function of ξ, $V_{\phi^2}(fg)(x,\xi)$ is simply the convolution $V_\phi f(x, \cdot) *$ $V_\phi g(x, \cdot)(\xi)$. Now, since $\langle \xi \rangle^s \lesssim \langle \xi - \eta \rangle^s + \langle \eta \rangle^s$, an application of Hölder and Young's inequalities gives first

$$\|fg\|_{\mathscr{M}_s^{p,q}} \lesssim \|f\|_{\mathscr{M}^{\infty,1}} \|g\|_{\mathscr{M}_s^{p,q}} + \|f\|_{\mathscr{M}_s^{p,q}} \|g\|_{\mathscr{M}^{\infty,1}}.$$

Finally, the embedding (e) of Proposition 5.3 finishes the proof.   □

We note that also that, in fact, the modulation spaces $\mathscr{M}_s^{p,q}(\mathbb{R}^d)$ are multiplication algebras *if and only if* $s > d/q'$; see [138, Theorem 1.5], also [164, Proposition B.1]. The same optimality result, that is, for $s > d/q'$, applies to weighted Fourier-Lebesgue spaces $\mathscr{F}L_s^p$ as well; see [164, Proposition B.2].

5.3 Nonlinear Operations on Weighted Modulation Spaces

Corollary 4.21 and the multilinear estimate (4.16) immediately following it have counterparts on weighted modulation spaces and the arguments proving them are slight variations of the arguments proving the unweighted results; see [14] and [141]. Specifically, we have that if $\sigma \in \mathscr{M}_s^{\infty,1}(\mathbb{R}^{(m+1)d})$ and $s \geq 0$, then the k-linear pseudodifferential operator L_σ extends to a bounded operator from $\mathscr{M}_s^{p_1,q_1} \times \cdots \times \mathscr{M}_s^{p_k,q_k}$ into $\mathscr{M}_s^{p_0,q_0}$ when $\frac{1}{p_1} + \cdots + \frac{1}{p_k} = \frac{1}{p_0}$, $\frac{1}{q_1} + \cdots + \frac{1}{q_k} = k - 1 + \frac{1}{q_0}$, and $0 < p_i \leq \infty, 1 \leq q_i \leq \infty$ for $0 \leq i \leq k$. In particular, we also obtain the weighted multilinear estimate

$$\left\| \prod_{i=1}^k u_i \right\|_{\mathscr{M}_s^{p_0,q_0}} \lesssim \prod_{i=1}^k \|u_i\|_{\mathscr{M}_s^{p_i,q_i}}, \tag{5.2}$$

with the same exponents as above. Moreover, using the embedding $\mathscr{M}_s^{p,1} \subseteq \mathscr{M}_s^{kp,1}$, we see that if $u \in \mathscr{M}_s^{p,1}$, then $u^k \in \mathscr{M}_s^{p,1}$ as well and

$$\|u^k\|_{\mathscr{M}_s^{p,1}} \lesssim \|u\|_{\mathscr{M}_s^{p,1}}^k. \tag{5.3}$$

We have in effect obtained a very simple instance of good behavior of a weighted modulation space under a *nonlinear operation*. Namely, if $F(z) = z^k$, and $F(u)$

denotes the composition of F with the function u, we have obtained the following statement:

$$\text{If } u \in \mathscr{M}_s^{p,1}, \text{ then } F(u) \in \mathscr{M}_s^{p,1}.$$

A natural question is whether this phenomenon occurs for other nonlinear transformations F as well. Indeed, when one is interested in the well-posedness theory of nonlinear partial differential equations with initial data in a modulation space, understanding the behavior of the modulation space under the nonlinearity becomes a necessity; see Chap. 7.

This type of question was asked before in the context of Bessel potential spaces L_s^p and answered in the positive by Bony [34] if F is sufficiently smooth and $F(0) = 0$. His argument relies on the theory of paradifferential operators and has a nice connection with the Hörmander classes of pseudodifferential symbols introduced in the previous chapter. This observation was then extended to include the larger family of Besov spaces; see [180] and [201].

Theorem 5.5 *Let $1 \leq p \leq \infty$ and $s > d/p$. Assume that $F \in \mathscr{C}^{\lceil s+1 \rceil}(\mathbb{R})$ and $F(0) = 0$. Then, for all real-valued functions $u \in L_s^p(\mathbb{R}^d)$ we have $F(u) \in L_s^p(\mathbb{R}^d)$ as well. Moreover, the following estimate holds:*

$$\|F(u)\|_{L_s^p} \lesssim \|F'\|_{\mathscr{C}^{\lceil s \rceil}} (1 + \|u\|_{L^\infty}^{\lceil s \rceil}) \|u\|_{L_s^p}.$$

Proof We only sketch the steps. One writes $F(u) = L_\sigma(u)$, where L_σ is a linear pseudodifferential operator with symbol $\sigma(x, \xi) = \sigma(x, \xi; F)$. If we further assume that $u \in L^\infty$, then $\sigma \in S_{1,1}^0$, which guarantees the boundedness of L_σ on L_s^p with $s > 0$; see [208, Proposition 5, p. 251]. The further assumption $s > d/p$ implies the embedding $L_s^p(\mathbb{R}^d) \subset L^\infty(\mathbb{R}^d)$. For the quantitative estimate on $\|F(u)\|_{L_s^p}$, see for example, [218, Sections 2.4 and 2.5] or [217, Section 3.1]. $\square$

Returning now to modulation spaces, it is tempting to infer that the pseudodifferential approach will again work in this context. However, as pointed out in Theorems 4.14 and 4.15, the modulation spaces are ill suited for the Hörmander classes of symbols. Nevertheless, we have the following analog of Theorem 5.5, see Sugimoto et al. [214, Theorem 4.2].

Theorem 5.6 *Let $s > d/2$, $k \in \mathbb{N}$ and $1 \leq p \leq 2(k + 1)$. Assume that $F \in \mathscr{C}^{k+\lceil s+1 \rceil}(\mathbb{R})$ and $F^{(j)}(0) = 0$ for $0 \leq j \leq k$. Then, for all real-valued $u \in \mathscr{M}_s^{p,2}(\mathbb{R}^d)$ we have $F(u) \in \mathscr{M}_s^{p,2}(\mathbb{R}^d)$ as well. Moreover, the following estimate holds:*

$$\|F(u)\|_{\mathscr{M}_s^{p,2}} \lesssim (1 + \|u\|_{\mathscr{M}_s^{p,2}}^{\lceil s \rceil}) \|u\|_{\mathscr{M}_s^{p,2}}^{k+1}.$$

Proof Let ϕ, ψ be two real-valued functions in $\mathscr{C}_c^\infty(\mathbb{R}^d)$ with $\psi(x) = 1$ for $x \in \text{supp}(\phi)$. Because the derivatives of F at 0 vanish up to order k, we have $F(u) = u^k G(u)$ for some $G \in \mathscr{C}^{\lceil s+1 \rceil}(\mathbb{R})$ and $G(0) = 0$. We note immediately that the last condition on G is exactly the one appearing in Theorem 5.5. Observe also that the assumptions on the windows ϕ, ψ imply in particular that for any fixed $x \in \mathbb{R}^d$, if $y \in \mathbb{R}^d$ is such that $y - x \in \text{supp}(\phi)$ then, $\psi(y - x) = 1$. Consequently, we can write

$$V_\phi(F(u))(x, \xi) = \int_{\mathbb{R}^d} \phi(y - x)\psi^k(y - x)u^k(y)G(\psi(y - x)u(y))e^{-2\pi i y \cdot \xi} \, dy.$$

Recall now that $\mathscr{M}_s^{p,q}(\mathbb{R}^d) \subset \mathscr{M}^{\infty,1}(\mathbb{R}^d) \subset L^\infty(\mathbb{R}^d)$ if $s > d/q'$; in particular, if $s > d/2$, we have $\mathscr{M}_s^{p,2}(\mathbb{R}^d) \subset L^\infty(\mathbb{R}^d)$. Therefore, using Propositions 5.5 and 1.63, we have the following

$$\|V_\phi(F(u))(x, \xi)\langle\xi\rangle^s\|_{L^2(d\xi)} \lesssim \|\phi(y - x)\psi^k(y - x)u^k(y)G(\psi(y - x)u(y))\|_{H^s(dy)}$$

$$\lesssim \|\phi(\cdot - x)\|_{H^s}\|\psi(\cdot - x)u(\cdot)\|_{H^s}^k\|G(\psi(\cdot - x)u(\cdot))\|_{H^s}$$

$$\lesssim_{\phi,\psi} (1 + \|u\|_{L^\infty})^{\lceil s \rceil}\|\psi(\cdot - x)u(\cdot)\|_{H^s}^{k+1}$$

$$\lesssim (1 + \|u\|_{\mathscr{M}_s^{p,2}})^{\lceil s \rceil}\|V_\psi u(x, \xi)\langle\xi\rangle^s\|_{L^2(d\xi)}^{k+1}.$$

Let now $1 \leq p \leq 2$. Using Minkowski's inequality (Theorem 1.3) and the embedding $\mathscr{M}_s^{p,2} \subset \mathscr{M}_s^{p(k+1),2}$, we can write:

$$\|F(u)\|_{\mathscr{M}_s^{p,2}} = \|V_\phi(F(u))(x, \xi)\langle\xi\rangle^s\|_{L^p(dx)L^2(d\xi)} \leq \|V_\phi(F(u))(x, \xi)\langle\xi\rangle^s\|_{L^2(d\xi)L^p(dx)}$$

$$\lesssim (1 + \|u\|_{\mathscr{M}_s^{p,2}})^{\lceil s \rceil}\|V_\psi u(x, \xi)\langle\xi\rangle^s\|_{L^2(d\xi)L^{p(k+1)}(dx)}^{k+1}$$

$$\lesssim (1 + \|u\|_{\mathscr{M}_s^{p,2}})^{\lceil s \rceil}\|V_\psi u(x, \xi)\langle\xi\rangle^s\|_{L^{p(k+1)}(dx)L^2(d\xi)}^{k+1}$$

$$= (1 + \|u\|_{\mathscr{M}_s^{p,2}})^{\lceil s \rceil}\|u\|_{\mathscr{M}_s^{p(k+1),2}}^{k+1} \lesssim (1 + \|u\|_{\mathscr{M}_s^{p,2}}^{\lceil s \rceil})\|u\|_{\mathscr{M}_s^{p,2}}^{k+1}.$$

In the case $2 \leq p \leq 2(k + 1)$, we first write $\|F(u)\|_{\mathscr{M}_s^{p,2}} \lesssim \|F(u)\|_{\mathscr{M}_s^{2,2}}$ and then proceed analogously as in the previous case with $p = 2$.   □

Let $\alpha > 0$ and consider the following two functions: $F_1(x) = |x|^\alpha$ and $F_2(x) = |x|^{\alpha-1}x$. Clearly, if α is an odd natural number, then $F_1 \in \mathscr{C}^{\alpha-1}$, while if α is an even natural number, then $F_2 \in \mathscr{C}^{\alpha-1}$. When α is a positive number that is not odd, respectively even, it is easy to check that $F_1 \in \mathscr{C}^{\lfloor\alpha\rfloor}$ and $F_2 \in \mathscr{C}^{\lfloor\alpha\rfloor}$, respectively. Moreover, $F_i^{(j)}(0) = 0$ for $1 \leq i \leq 2, 0 \leq j \leq \lfloor\alpha\rfloor$. Thus, letting $\lfloor\alpha\rfloor > \lfloor s \rfloor + 2$ in Theorem 5.6, we obtain the following.

Corollary 5.7 *Let $s > d/2$, $\alpha > 0$ be a non-integer such that $\lfloor\alpha\rfloor > \lfloor s \rfloor + 2$ and $1 \leq p \leq 2(\lfloor\alpha\rfloor - \lfloor s \rfloor - 1)$. If $u \in \mathscr{M}_s^{p,2}(\mathbb{R}^d)$ is real-valued, then $|u|^\alpha$, $|u|^{\alpha-1}u \in \mathscr{M}_s^{p,2}(\mathbb{R}^d)$ as well.*

For a related work on non-integer power estimates on modulation spaces, see also [50]. The previous corollary raises the hope that similar results may work for less restrictive conditions on the α power in the nonlinearity and to include more of the spectrum of modulation spaces; but strictly based on the previous calculations, one can also extrapolate that answering this question affirmatively for general power-type nonlinearities of the form $|u|^\alpha u$ may not be possible.

5.4 Operating Functions on Modulation Spaces

Corollary 5.7 of the previous section can be recast into a statement about the nonlinear functions F_1, F_2 *operating* on some of the modulation spaces $\mathscr{M}_s^{p,q}$ as long as the exponent α is sufficiently large. We have the following definition.

Definition 5.8 Let $F : \mathbb{R}^2 \to \mathbb{C}$. We say that F operates on $\mathscr{M}_s^{p,q}$ if for every (complex valued) $f \in \mathscr{M}_s^{p,q}$ we have $F(\text{Re } f, \text{Im } f) \in \mathscr{M}_s^{p,q}$.

Recent works of Bhimani-Ratnakumar [28, 29] and Kobayashi-Sato [168] have settled completely the question about those F's which operate on $\mathscr{M}_s^{p,q}$ when $q = 1$ and $s = 0$.

Theorem 5.9 *Let* $1 \le p < \infty$ *and* $F : \mathbb{R}^2 \to \mathbb{C}$. *Then* F *operates on* $\mathscr{M}^{p,1}(\mathbb{R})$ *if and only if* F *is a real analytic function with* $F(0) = 0$.

We recall that F is called *real analytic* if for each $(x_0, y_0) \in \mathbb{R}^2$, we can write

$$F(x, y) = \sum_{m=0}^{\infty} \sum_{n=0}^{\infty} c_{mn}(x - x_0)^m (y - y_0)^n$$

and this series converges absolutely in a neighborhood of (x_0, y_0). Let us also denote by $\mathscr{A}(\mathbb{T})$ the set of all continuous functions on the torus $\mathbb{T}$ with absolutely convergent Fourier series, that is

$$\|f\|_{\mathscr{A}(\mathbb{T})} := \sum_{n \in \mathbb{Z}} |\hat{f}(n)| < \infty,$$

where $\hat{f}(n) = \int_0^{2\pi} f(t)e^{-2\pi i n t}\, dt$. In what follows, we only indicate the basic ingredients and steps required in the proof of the "if" part of Theorem 5.9 by following closely [168].

Proof We want to show that if F is real analytic with $F(0) = 0$ and $f \in \mathscr{M}^{p,1} = \mathscr{M}^{p,1}(\mathbb{R})$, then $F(\text{Re } f, \text{Im } f) \in \mathscr{M}^{p,1}$. We consider separately the "local" and "global" behaviors of $G = F(\text{Re } f, \text{Im } f)$; specifically, for a given $\phi \in \mathscr{C}_c^\infty(\mathbb{R})$, by writing $G = \phi G + (1 - \phi)G$, it suffices to show that both the local component ϕG and global component $(1 - \phi)G$ belong to $\mathscr{M}^{p,1}$.

We begin with the local part. To fix ideas, let $z_0 \in \mathbb{R}$ and $\varphi \in \mathscr{C}_c^\infty(\mathbb{R})$ be so that supp $(\varphi) \in \mathbb{B}(z_0, 0.1)$ and $\varphi \equiv 1$ on $\mathbb{B}(z_0, 0.05)$. Let $f \in \mathscr{M}^{p,1}$ and write $f(z_0) = x_0 + iy_0$, with $x_0, y_0 \in \mathbb{R}$. Define f_1, f_2 as $f_1 = T_{x_0}(\mathrm{Re}\, f)$ and $f_2 = T_{y_0}(\mathrm{Im}\, f)$. Clearly, $(\varphi f_j)(z_0) = 0$ for $j = 1, 2$ and, due to (5.2) and the fact that $\varphi \in \mathscr{S} \subset \mathscr{M}^{p,1} \subset \mathscr{M}^{2p,1}$, we have

$$\|\varphi f_j\|_{\mathscr{M}^{p,1}} \lesssim \|\varphi\|_{\mathscr{M}^{2p,1}} \|f_j\|_{\mathscr{M}^{2p,1}} \lesssim \|\varphi\|_{\mathscr{M}^{p,1}} \|f_j\|_{\mathscr{M}^{p,1}} < \infty,$$

hence giving $\varphi f_j \in \mathscr{M}^{p,1}$ for $j = 1, 2$; in essence, we have reproved here that $\mathscr{M}^{p,1}$ is an algebra with respect to multiplication. By further using [8, Proposition B.1], the condition $\varphi \in \mathscr{C}_c^\infty$ gives $\|\varphi f_j\|_{\mathscr{A}(\mathbb{T})} \lesssim \|f_j\|_{\mathscr{M}^{p,1}} < \infty$, thus $\varphi f_j \in \mathscr{A}(\mathbb{T})$ for $j = 1, 2$.

Recall now that since F is real analytic, we can write

$$F(x_0, y_0) = \sum_{m=0}^{\infty} \sum_{n=0}^{\infty} c_{mn} (x - x_0)^m (y - y_0)^n$$

with an absolute convergent series for some $\delta > 0$ and $|x - x_0| + |y - y_0| < \delta$. Let now $\epsilon > 0$ be so that $\epsilon < \delta$. For $\lambda \in (0, 1)$ and $t \in \mathbb{T}$, denote by $V_\lambda(t) = 2\max\{0, 1 - |t|/(2\lambda)\} - \max\{0, 1 - |t|/\lambda\}$. It can be shown [158, pp. 56–57] that as $\lambda \to 0^+$, we have $\|(T_{z_0} V_\lambda)\varphi f_j\|_{\mathscr{A}(\mathbb{T})} \to 0$, thus there exists a $\lambda_0 > 0$ so that for $0 < \lambda < \lambda_0$, $\|(T_{z_0} V_\lambda)\varphi f_j\|_{\mathscr{A}(\mathbb{T})} \lesssim \epsilon$. Fix $\lambda \in (0, \lambda_0)$. By using again [8, Proposition B.1], we can now write $\|\varphi(T_{z_0} V_\lambda)\varphi f_j\|_{\mathscr{M}^{p,1}} \lesssim \|(T_{z_0} V_\lambda)\varphi f_j\|_{\mathscr{A}(\mathbb{T})} \lesssim \epsilon$.

Define $H = \displaystyle\sum_{m=0}^{\infty} \sum_{n=0}^{\infty} c_{mn} (\varphi(T_{z_0} V_\lambda)\varphi f_1)^m \varphi(T_{z_0} V_\lambda)\varphi f_2)^n$. Note that since φ is supported around z_0, we have that for $z \in \mathbb{B}(z_0, 0.05)$, $\varphi(T_{z_0} V_\lambda)\varphi f_j(z) = f_j(z)$; thus $H(z) = G(z)$ for $z \in \mathbb{B}(z_0, 0.05)$. Now, because $F(0) = 0$, we clearly have $c_{00} = 0$. Using once more the algebra property of $\mathscr{M}^{p,1}$, we get

$$\|H\|_{\mathscr{M}^{p,1}} \lesssim \sum_{m=0}^{\infty} \sum_{n=0}^{\infty} |c_{mn}| \|\varphi(T_{z_0} V_\lambda)\varphi f_1\|_{\mathscr{M}^{p,1}}^m \|\varphi(T_{z_0} V_\lambda)\varphi f_2\|_{\mathscr{M}^{p,1}}^n \qquad (5.4)$$

$$\lesssim \sum_{m=0}^{\infty} \sum_{n=0}^{\infty} |c_{mn}| \epsilon^{m+n} \lesssim \sum_{m=0}^{\infty} \sum_{n=0}^{\infty} |c_{mn}| \delta^{m+n} < \infty. \qquad (5.5)$$

Thus, $H \in \mathscr{M}^{p,1}$, showing that G belongs to $\mathscr{M}^{p,1}$ locally at any point z_0.

We consider next the global part. Using [168, Proposition 3.2], we see that for all $f \in \mathscr{M}^{p,1}$ and $\delta > 0$ as above, there exists $\psi \in \mathscr{C}_c^\infty(\mathbb{R})$ with supp $(\psi) \subset \mathbb{B}(0, 1)$ so that $\|(1 - \psi)\mathrm{Re}\, f\|_{\mathscr{M}^{p,1}} < \delta$ and $\|(1 - \psi)\mathrm{Im}\, f\|_{\mathscr{M}^{p,1}} < \delta$. Again, since F is real analytic at 0, similar considerations as above show that $K = \displaystyle\sum_{m=0}^{\infty} \sum_{n=0}^{\infty} c_{mn} ((1 -$

ψ)Re $f)^m((1-\psi)$Im $f)^n$ belongs to $\mathscr{M}^{p,1}$. Moreover, $K = G$ outside $\overline{\mathbb{B}(0,1)}$, thus proving that G belongs to $\mathscr{M}^{p,1}$ globally as well. The proof of the "if" implication is complete. $\square$

The argument in [28] in fact proves the case $p = 1$ of Theorem 5.9 for the Feichtinger algebra $\mathscr{M}^{1,1}(\mathbb{R}^d)$ for all $d \geq 1$. Thus, naturally, one can further ask about other (nonlinear) functions operating on more general modulation spaces $\mathscr{M}_s^{p,q}(\mathbb{R}^d)$. We note that since $\mathscr{M}_s^{p,q}(\mathbb{R}^d)$ is a multiplication algebra for $s > d/q'$ (see Proposition 5.4), it immediately follows that as long as $F(z)$ is an entire function, and $f \in \mathscr{M}_{d/q'+\epsilon}^{p,q}(\mathbb{R}^d)$ we also have $F(f) \in \mathscr{M}_{d/q'+\epsilon}^{p,q}(\mathbb{R}^d)$ for all $\epsilon > 0$; see again also the discussion at the beginning of Sect. 5.3.

Interestingly, this operating behavior of some nonanalytic functions F on $\mathscr{M}_{d/q'+\epsilon}^{p,q}(\mathbb{R}^d)$ continues to hold as long as $q \in [4/3, \infty)$. The following result is due to Kato et al. [164, Theorem 1.1].

Theorem 5.10 Let $1 \leq p < \infty, 4/3 \leq q < \infty$ and $s > d/q'$. If $f : \mathbb{R}^d \to \mathbb{R}$, $f \in \mathscr{M}_s^{p,q}(\mathbb{R}^d)$, $F \in \mathscr{C}^\infty(\mathbb{R})$ and $F(0) = 0$, then $F(f) \in \mathscr{M}_s^{p,q}(\mathbb{R}^d)$.

Proof For $F \in \mathscr{C}^\infty$ with $F(0) = 0$, we can write for all $N \in \mathbb{N}$:

$$F(f) = G(f) + \sum_{k=1}^{N} \frac{F^{(k)}(0)}{k!} f^k, \qquad (5.6)$$

where $G(0) = G'(0) = \cdots = G^{(N)}(0) = 0$. Furthermore, by applying a Taylor expansion to G, we can also write

$$G(f) = f^N H(f), \quad \text{with } H \in C^\infty(\mathbb{R}), \ H(0) = 0. \qquad (5.7)$$

$\square$

Let $f \in \mathscr{M}_s^{p,q}$ with p, q, s as in the statement of our theorem. Since, by Proposition 5.4, $\mathscr{M}_s^{p,q}$ with $s > d/q'$ is a multiplication algebra, we have

$$\left\| \sum_{k=1}^{N} \frac{F^{(k)}(0)}{k!} f^k \right\|_{\mathscr{M}_s^{p,q}} \lesssim \sum_{k=1}^{N} \|f\|_{\mathscr{M}_s^{p,q}}^k.$$

Thus, it suffices to show that $G(f) \in \mathscr{M}_s^{p,q}$. Let us fix N for the moment and suppose that $\phi, \psi \in \mathscr{C}_c^\infty(\mathbb{R})$ are such that $\psi \equiv 1$ on supp (ϕ). For $x, \xi \in \mathbb{R}^d$ we have

$$V_\phi[G(f)](x, \xi) = \int_{\mathbb{R}^d} e^{-2\pi i \xi \cdot y} \phi(y - x) G(\psi(y - x) f(y)) \, dy$$

$$= \int_{\mathbb{R}^d} e^{-2\pi i \xi \cdot y} \phi(y - x)(\psi(y - x) f(y))^N H(\psi(y - x) f(y)) \, dy$$

$$= \mathscr{F}[T_x \phi (f T_x \psi)^N H(f T_x \psi)](\xi).$$

Multiplying on both sides of the equality above by the weight $\sigma_s(\xi) = \langle \xi \rangle^s$ and using the fact that $\mathscr{F}L_s^q$ is a multiplication algebra for $s > d/q'$, we have

$$\|\sigma_s(\xi)V_\phi[G(f)](x,\xi)\|_{L^q(d\xi)} = \left\|\left\|\sigma_s\mathscr{F}[T_x\phi(fT_x\psi)^N H(fT_x\psi)]\right\|\right\|_{L^q}$$
$$\lesssim \|T_x\phi\|_{\mathscr{F}L_s^q}\|fT_x\psi\|_{\mathscr{F}L_s^q}^N\|H(fT_x\psi)\|_{\mathscr{F}L_s^q}.$$

Now, observe that $\|T_x\phi\|_{\mathscr{F}L_s^q} \sim 1$ and $\|fT_x\psi\|_{\mathscr{F}L_s^q} = \|\sigma_s(\xi)V_\psi f(x,\xi)\|_{L^q(d\xi)}$. Moreover, by Proposition 2.9, which applies verbatim to the case of weighted modulation spaces, see also [164, Lemma A.1], and the fact that $\mathscr{M}_s^{p,q}$ is a multiplication algebra, we immediately get

$$\|fT_x\psi\|_{\mathscr{F}L_s^q} \sim \|fT_x\psi\|_{\mathscr{M}_s^{p,q}} \lesssim \|\psi\|_{\mathscr{M}_s^{p,q}}\|f\|_{\mathscr{M}_s^{p,q}},$$

with the implicit constants independent of x. Moreover, it can be shown [164, Proposition 4.1, Theorem 1.2] that

$$\sup_{x\in\mathbb{R}^d} \|H(fT_x\psi)\|_{\mathscr{F}L_s^q} < \infty,$$

which in turn yields the following estimate:

$$\|\sigma_s V_\phi[G(f)](x,\cdot)\|_{L^q} \lesssim \|\sigma_s V_\psi f(x,\cdot)\|_{L^q}^M, \qquad (5.8)$$

with the implicit constant independent of $x \in \mathbb{R}^d$ and with $M > N$ arbitrarily large.

Assume now that $p \le q$. Using Minkowski's integral inequality and (5.8) we can write

$$\|G(f)\|_{\mathscr{M}_s^{p,q}} \lesssim \left\|\left\|\sigma_s(\xi)V_\phi[G(f)](x,\xi)\|_{L^q(d\xi)}\right\|\right\|_{L^p(dx)}$$
$$\lesssim \left\|\left\|\sigma_s(\xi)V_\psi f(x,\xi)\|_{L^q(d\xi)}^M\right\|\right\|_{L^p(dx)}$$
$$= \left\|\left\|\sigma_s(\xi)V_\psi f(x,\xi)\|_{L^q(d\xi)}\right\|\right\|_{L^{Mp}(dx)}^M.$$

Choose now $M > \lceil \max(p/q, q/p) \rceil$, so that $Mp > q \ge p$. By Proposition 5.3 (d), we can then further write

$$\|G(f)\|_{\mathscr{M}_s^{p,q}} \lesssim \left\|\left\|\sigma_s(\xi)V_\psi f(x,\xi)\|_{L^{Mp}(dx)}\right\|\right\|_{L^q(d\xi)}^M$$
$$\lesssim \left\|\left\|\sigma_s(\xi)V_\psi f(x,\xi)\|_{L^p(dx)}\right\|\right\|_{L^q(d\xi)}^M \sim \|f\|_{\mathscr{M}_s^{p,q}}^M.$$

A similar argument applies to the case $q < p < \infty$, thus proving

$$\|G(f)\|_{\mathscr{M}_s^{p,q}} \lesssim \|f\|_{\mathscr{M}_s^{p,q}}^M.$$

The proof is complete.   □

5.5 Notes

Weighted modulation spaces can be defined in greater generality. Given a positive *submultiplicative weight* v on $\mathbb{R}^{2d}$, that is, a function that satisfies $v(z_1 + z_2) \le v(z_1)v(z_2)$ for all $z_1, z_2 \in \mathbb{R}^{2d}$, we say that ω is a v-*moderate weight* if ω is a positive function on $\mathbb{R}^{2d}$ and there is some constant $C > 0$ such that $\omega(z_1 + z_2) \le Cv(z_1)\omega(z_2)$. Now, given a nonzero window $\phi \in \mathscr{S}(\mathbb{R}^d)$ and ω a v-moderate function, where we assume that v has a polynomial growth, the modulation space $\mathscr{M}_\omega^{p,q}(\mathbb{R}^d)$ consists of all $f \in \mathscr{S}'(\mathbb{R}^d)$ such that $V_\phi f \in L_\omega^{p,q}$, with the appropriate norm

$$\|f\|_{\mathscr{M}_\omega^{p,q}} := \left(\int_{\mathbb{R}^d} \left(\int_{\mathbb{R}^d} |V_\phi f(x,\xi)|^p \omega(x,\xi)^p \, dx \right)^{q/p} d\xi \right)^{1/q}$$

and the usual changes for $p = \infty$ or $q = \infty$. In this context, the spaces $\mathscr{M}_s^{p,q}$ discussed in this chapter are simply $\mathscr{M}_{\sigma_s}^{p,q}$; note that $\sigma_{|s|}$ is submultiplicative and σ_s is a $\sigma_{|s|}$-moderate weight. For more on "good" weights for time-frequency analysis and properties of such weighted modulation spaces, see again [126] and [125].

The Wiener algebra property of certain symbol classes of pseudodifferential operators combined with other pseudodifferential calculus techniques can be used to construct some explicit isomorphisms between more general weighted modulation spaces with appropriate multiplicative weights, thus showing that the whole family of modulation spaces possesses the lifting property, see [131].

Several continuity results for pseudodifferential operators on modulation spaces can be extended to weighted modulation spaces. Simply to give a flavor of what such generalizations would look like, we state two results of Toft [219, Theorem 3.2] and [222, Theorem 3.2'].

Theorem 5.11 *Assume that v is some polynomial on $\mathbb{R}^d$, ω_1 is a v-moderate weight, and ω_2 is a smooth v-moderate weight such that $(\partial^\alpha \omega_2)/\omega_2 \in L^\infty(\mathbb{R}^d)$ for all indices α. Let $1 \le p, q \le \infty$ and consider $\omega_2(x, \xi) = \omega_2(x)$ or $\omega_2(x, \xi) = \omega_2(\xi)$. Then the pseudodifferential operator $\omega_2(x, D)$ is bounded from $\mathscr{M}_{\omega_1 \omega_2}^{p,q}(\mathbb{R}^d)$ to $\mathscr{M}_{\omega_1}^{p,q}(\mathbb{R}^d)$.*

Theorem 5.12 *Assume that v is some polynomial on $\mathbb{R}^d$, ω_1, ω_2 are v-moderate weights and σ is smooth such that $\partial^\alpha \sigma/\omega_2 \in L^\infty$ for all indices α. Then, for all $1 \le p, q \le \infty$, $t \in \mathbb{R}$, the t-pseudodifferential operator $\sigma_t(x, D)$ is bounded from $\mathscr{M}_{\omega_1 \omega_2}^{p,q}(\mathbb{R}^d)$ to $\mathscr{M}_{\omega_2}^{p,q}(\mathbb{R}^d)$.*

Chapter 6
Modulation Spaces and Other Function Spaces

We have by now defined both the weighted and unweighted versions of the modulation spaces and indicated in Chaps. 4 and 5 their place in the time-frequency analysis of pseudodifferential operators and their good properties regarding certain nonlinear operations stemming from PDEs. Hidden in the seemingly simple definition of modulation spaces lies the practical problem of computing the STFT of a function or distribution and then further calculating its appropriate mixed Lebesgue norm. Therefore, a natural question arising in this context is whether any *embeddings* between such modulation spaces and other classical function spaces of analysis exist, something that we have alluded to already in Sects. 2.3 and 2.4. Let us recall also immediately that Sjöstrand's class, $\mathcal{M}^{\infty,1}$, contains the Hörmander class $S^0_{0,0}$, a fact that played a relevant role throughout Chap. 4; see again the comments after Corollary 4.22 and [206].

There is a plethora of other applications to which the issue of an appropriate embedding applies. By changing the L^2-norm in the lower bound of the Heisenberg-Pauli-Weyl inequality

$$\|xf(x)\|^2_{L^2} + \|\xi\,\widehat{f}(\xi)\|^2_{L^2} \geq \frac{1}{2\pi}\|f\|^2_{L^2}$$

with natural measures of the time-frequency concentration of f reflected in its STFT, Galperin-Gröcheing [113] derived a whole class of uncertainty principles that are effectively equivalent to optimal embeddings of weighted spaces of the form

$$L^p(\sigma_a) \cap \mathscr{F}L^{b,q} \subset \mathcal{M}^{r,s}_{\sigma_\alpha \otimes \sigma_\beta},$$

for appropriate parameters $a, b, \alpha, \beta, p, q, r, s$. These type of embeddings can also be interpreted as extensions of the Cowling-Price uncertainty principles in [84], while the particular case $r = s = 1$, $\alpha = \beta = 0$ is related to Poisson's summation formula and has applications to pseudodifferential operators [124]; see also [112] for a criterion of compactness of this embedding that relies on the tightness of the STFT

© Springer Science+Business Media, LLC, part of Springer Nature 2020
Á. Bényi, K. A. Okoudjou, *Modulation Spaces*, Applied and Numerical
Harmonic Analysis, https://doi.org/10.1007/978-1-0716-0332-1_6

[91, Theorem 4] applicable to the more general context of co-orbit spaces, as well as [149] for some of the critical cases explored in [124]. Spurred by a result of Gautam [116] that shows a neat connection between the VMO space of vanishing mean oscillations and the Balian-Low theorem, fundamental to the theory of Gabor frames and which is stated in the setting of weighted Sobolev spaces, Heil-Tinaztepe [144] investigated the modulation space characterization of the Zak transform and were naturally led to consider the embeddings of modulation spaces into the BMO space of bounded mean oscillations; both the BMO and VMO spaces are central function spaces in harmonic analysis, where they appear prominently in investigations of boundedness and compactness criteria for singular integral operators.

Perhaps one of the simplest and most glaring examples of such embeddings between modulation spaces and other classical function spaces encoding smoothness is already present in a particular case of the Sobolev embedding theorem, namely $H^s(\mathbb{R}^d) \subset \mathscr{C}^k(\mathbb{R}^d)$ for $s > k + d/2$; see also Proposition 1.61. Indeed, H^s can be identified both with the modulation space $\mathscr{M}_s^{2,2}$ and Besov space $B_s^{2,2}$, while $\mathscr{C}^k$ is a Besov space. In general, these two classes of function spaces, modulation and Besov, are distinct for parameters $p, q \neq 2$. Given the ubiquitous nature of the Besov and Bessel potential spaces as smoothness spaces for the initial profile of various PDEs, the question of how they embed into modulation spaces and vice versa is a natural one and it features prominently in the first systematic studies of these topics by Okoudjou [193] and Toft [219–221], but goes back as far as Gröbner's unpublished work [123] on α-modulation spaces; see [223] and [139] for more on the embedding properties of the latter spaces.

In this chapter we discuss the comparison between the modulation spaces and the family of Besov spaces by indicating the (sharp) embeddings that hold between them. Furthermore, we will present the optimal embeddings between the modulation spaces and the Bessel potential (or L^p-Sobolev) spaces.

6.1   Embeddings Between Modulation Spaces and Besov Spaces

We start first with the definition of inhomogeneous Besov spaces and some of their basic properties. Let $\varphi_0, \varphi \in \mathscr{S}(\mathbb{R}^d)$ such that $\text{supp}(\varphi_0) \subset \{|\xi| \leq 2\}$, $\text{supp}(\varphi) \subset \{\frac{1}{2} \leq |\xi| \leq 2\}$, and $\varphi_0(\xi) + \sum_{j=1}^{\infty} \varphi(2^{-j}\xi) \equiv 1$. With $\varphi_j(\xi) = \varphi(2^{-j}\xi)$, $j \geq 1$, we define *the (inhomogeneous) Besov space* $B_s^{p,q} := B_s^{p,q}(\mathbb{R}^d)$ via the (quasi) norm

$$\|f\|_{B_s^{p,q}} = \|(2^{sj} f * \mathscr{F}^{-1}(\varphi_j))_{j \geq 0}\|_{L^p \ell^q} = \left(\sum_{j \geq 0} 2^{sjq} \|f * \mathscr{F}^{-1}(\varphi_j)\|_{L^p}^q \right)^{1/q};$$

(6.1)

when $q = \infty$, the norm is $\sup_{j \geq 0} 2^{sj} \|f * \mathscr{F}^{-1}(\varphi_j)\|_{L^p}$. We point out that, using the notation for multiplier (or, more generally, pseudodifferential) operators from Chap. 4,

$$\sigma(D)f := L_\sigma(f) = f * \sigma^\vee,$$

we can also write

$$\|f\|_{B_s^{p,q}} = \|(2^{sj}\varphi_j(D)f)_{j\geq0}\|_{L^p\ell^q}. \tag{6.2}$$

One can define the homogeneous version of these spaces by an appropriate modification of the quasi-norm above. Note also that since $\varphi_j = 2^{jd}D_{2^j}\varphi$, we also have $\mathscr{F}^{-1}(\varphi_j) = D_{2^{-j}}(\varphi^\vee)$. As expected, the definition of $B_s^{p,q}$ is independent of the functions φ_0 and φ above, different choices leading to equivalent norms; for $0 < p, q < \infty$, $\mathscr{S}(\mathbb{R}^d)$ is a dense subset of all $B_s^{p,q}(\mathbb{R}^d)$ and $(B_s^{p,q})' = B_{-s}^{p',q'}$; and, if $0 < q_1 \leq q_2 \leq \infty$, $0 < p \leq \infty$ and $s_1 < s_2$, then we have the following embeddings: $B_s^{p,q_1} \subset B_s^{p,q_2}$, $B_{s_2}^{p,\infty} \subset B_{s_1}^{p,1}$. Like the Bessel potential spaces and the modulation spaces, the Besov spaces interpolate as expected and for appropriate indices s they are multiplication algebras. The following results are proved in [227, Theorem 2.4.7] and [227, Theorem 2.8.3].

Proposition 6.1 Let $1 \leq p_1, q_1, p_2, q_2 \leq \infty$, $s_1, s_2 \in \mathbb{R}$, $0 \leq \theta \leq 1$. Define

$$s = \theta s_1 + (1-\theta)s_2, \frac{1}{p} = \frac{\theta}{p_1} + \frac{1-\theta}{p_2}, \frac{1}{q} = \frac{\theta}{q_1} + \frac{1-\theta}{q_2}.$$

Then, if $f \in B_{s_1}^{p_1,q_1} \cap B_{s_2}^{p_2,q_2}$, we have $f \in B_s^{p,q}$ and

$$\|f\|_{B_s^{p,q}} \lesssim \|f\|_{B_{s_1}^{p_1,q_1}} \|f\|_{B_{s_2}^{p_2,q_2}}.$$

Proposition 6.2 Let $1 \leq p, q \leq \infty$ and $s > d/p$. Then, for all $f, g \in B_s^{p,q}(\mathbb{R}^d)$ we have

$$\|fg\|_{B_s^{p,q}} \lesssim \|f\|_{B_s^{p,q}} \|f\|_{B_s^{p,q}}.$$

Recall now that in Chap. 2 we have defined the indices $\mu_1(p,q)$ and $\mu_2(p,q)$ that were essential in establishing the dilation property of modulation spaces; see (2.4) and (2.5). These yield in turn the indices $\theta_i = \mu_i + 1/p$, $i = 1, 2$. The explicit expressions of θ_1 and θ_2 are the following:

$$\theta_1(p,q) = \begin{cases} 0 & \text{if } (1/p, 1/q) \in \mathbb{I}_1^*, \\ 1/p + 1/q - 1 & \text{if } (1/p, 1/q) \in \mathbb{I}_2^*, \\ -1/p + 1/q & \text{if } (1/p, 1/q) \in \mathbb{I}_3^*, \end{cases} \tag{6.3}$$

and

$$\theta_2(p,q) = \begin{cases} 0 & \text{if } (1/p, 1/q) \in \mathbb{I}_1, \\ 1/p + 1/q - 1 & \text{if } (1/p, 1/q) \in \mathbb{I}_2, \\ -1/p + 1/q & \text{if } (1/p, 1/q) \in \mathbb{I}_3; \end{cases} \tag{6.4}$$

for the definition of the regions $\mathbb{I}_j$ and $\mathbb{I}_j^*$, $j = 1, 2, 3$, see again Chap. 2. It turns out that, for $1 \le p, q \le \infty$, we have the identities

$$\theta_1(p,q) = \max\left(0, 1/q - \min(1/p, 1/p')\right),$$

$$\theta_2(p,q) = \min\left(0, 1/q - \max(1/p, 1/p')\right).$$

In particular, any of the functions θ_1, θ_2 determines the other simply because $\theta_1(p,q) = -\theta_2(p', q')$. With this notation, Toft proved in [220, Theorem 3.1] the following result.

Theorem 6.3 *For $1 \le p, q \le \infty$, we have the embeddings*

$$B_{d\theta_1(p,q)}^{p,q}(\mathbb{R}^d) \subset \mathscr{M}^{p,q}(\mathbb{R}^d) \subset B_{d\theta_2(p,q)}^{p,q}(\mathbb{R}^d).$$

More precisely, there exists a constant $C = C(d) > 0$ such that for all $f \in \mathscr{S}(\mathbb{R}^d)$ we have

$$C^{-1}\|f\|_{B_{d\theta_2(p,q)}^{p,q}} \le \|f\|_{\mathscr{M}^{p,q}} \le C\|f\|_{B_{d\theta_1(p,q)}^{p,q}}.$$

The proof of Theorem 6.3 relies upon the next result.

Lemma 6.4 *Let $1 \le p \le \infty$. Then $\mathscr{M}^{p,1} \subset B_0^{p,1}$ and $B_0^{p,\infty} \subset \mathscr{M}^{p,\infty}$.*

Proof Note that by duality it suffices to prove $B_0^{p,\infty} \subset \mathscr{M}^{p,\infty}$. Suppose that $f \in B_0^{p,\infty}$, $\phi \in \mathscr{C}_c^\infty$ is a nonzero window function and $N \in \mathbb{N}$ is a sufficiently large integer. With the notation in (6.1), there exists a $k \in \mathbb{N}$ such that for all $\eta \in \text{supp}(T_\xi \phi)$, we have

$$\sum_{j=0}^{N} \varphi_{k+j}(\eta) = 1.$$

Recall also that we defined the STFT of f with respect to ϕ as

$$V_\phi f(x, \xi) = \mathscr{F}(f\overline{T_x \phi})(\xi)$$

and that, by the fundamental identity of STFT (see Proposition 1.82), we know $|V_{\hat\phi} f(x, \xi)| = |V_\phi \hat f(\xi, x)|$. Therefore,

$$\left\|V_{\hat{\phi}}f(x,\xi)\right\|_{L_x^p} = \left\|\mathscr{F}(\widehat{f T_{\xi}\phi})(x)\right\|_{L_x^p} \leq \sum_{j=0}^{N} \left\|\mathscr{F}(\varphi_{k+j}\widehat{f T_{\xi}\phi})(x)\right\|_{L_x^p}.$$

By further using Parseval's identity (see Proposition 1.51) and Young's inequality (see Theorem 1.11), we get

$$\left\|\mathscr{F}(\varphi_{k+j}\widehat{f T_{\xi}\phi})(x)\right\|_{L_x^p} \leq \left\||\widehat{\varphi_{k+j}\widehat{f}}| * |\widehat{\phi}|\right\|_{L^p}$$

$$\leq \left\|\widehat{\varphi_{k+j}\widehat{f}}\right\|_{L^p}\left\|\widehat{\phi}\right\|_{L^1} \lesssim \|f\|_{B_0^{p,\infty}}.$$

Thus,

$$\left\|V_{\hat{\phi}}f(x,\xi)\right\|_{L_x^p} \lesssim \|f\|_{B_0^{p,\infty}},$$

which clearly implies that

$$\|f\|_{\mathscr{M}^{p,\infty}} = \left\|V_{\hat{\phi}}f(x,\xi)\right\|_{L_x^p L_\xi^\infty} \lesssim \|f\|_{B_0^{p,\infty}}.$$

□

Due to the nested property of modulation spaces, Theorem 6.3 has the following immediate consequence.

Corollary 6.5 *Assume that* $1 \leq p, q, p_i, q_i \leq \infty, i = 1, 2,$ *satisfy* $p_1 \leq p \leq p_2$ *and* $q_1 \leq q \leq q_2$. *Then*

$$B_{d\theta_1(p_1,q_1)}^{p_1,q_1}(\mathbb{R}^d) \subset \mathscr{M}^{p,q}(\mathbb{R}^d) \subset B_{d\theta_2(p_2,q_2)}^{p_2,q_2}(\mathbb{R}^d).$$

We point out that the inclusion relation between modulation spaces and Besov spaces in Theorem 6.3 is optimal. The optimality result relies crucially on the dilation property of modulation spaces that we explored in Theorem 2.14. The optimality of the left embedding for $p = q \in [1, 2]$, respectively of the right embedding for $p = q \in [2, \infty]$, was already noticed in [220, Remark 3.11]; see also [154]. We have the following result of Sugimoto-Tomita [211, Theorem 1.2].

Theorem 6.6 *Let* $1 \leq p, q \leq \infty$ *and* $s \in \mathbb{R}$. *The following are true:*

(1) *If* $B_s^{p,q}(\mathbb{R}^d) \subset \mathscr{M}^{p,q}(\mathbb{R}^d)$, *then* $s \geq d\theta_1(p, q)$.
(2) *If* $p \neq \infty, q \neq \infty$ *and* $\mathscr{M}^{p,q}(\mathbb{R}^d) \subset B_s^{p,q}(\mathbb{R}^d)$, *then* $s \leq d\theta_2(p, q)$.

Sketch of Proof of Theorem 6.3 We start by noting that it is enough to prove (1) since (2) follows by duality and the fact that $\theta_2(p, q) = -\theta_1(p', q')$. Thus, it suffices to concentrate on pairs $(1/p, 1/q)$ which belong to one of the regions $\mathbb{I}_k^*, k = 1, 2, 3$. We only indicate the argument for $(1/p, 1/q) \in \mathbb{I}_1^*$. In this case,

$\theta_1(p, q) = 0$. Assume first that $(p, q) \neq (1, \infty)$. Suppose also, by contradiction, that $B_s^{p,q} \subset \mathcal{M}^{p,q}$ with $s = -\epsilon < 0$. For this fixed $\epsilon > 0$, construct the function

$$f(x) = e^{8ix_1} \sum_{j \neq 0} |j|^{-d/p-\epsilon/2} (T_j \psi^{\vee})(x), \tag{6.5}$$

where $x = (x_1, x_2, \ldots, x_d) \in \mathbb{R}^d$, and $\psi \in \mathscr{C}_c^{\infty}(\mathbb{R}^d)$ such that supp $(\psi) \subset \overline{\mathbb{B}(0, 1)}$ and $\psi \equiv 1$ on $\overline{\mathbb{B}(0, 1/2)}$. Recalling now the notation $f^{\lambda}(x) = f(\lambda x)$, one can obtain the following estimates for the dyadic dilations of the function f in (6.5); with $\lambda = 2^k, k \in \mathbb{N}$, we have

(a) $\|f^{\lambda}\|_{\mathcal{M}^{p,q}} \gtrsim \lambda^{-d/p-\epsilon/2}$, for all $k \geq \lceil (\log_2 d)/2 \rceil$;
(b) $\|f^{\lambda}\|_{B_s^{p,q}} \lesssim \lambda^{-d/p+s}$, for all $k \in \mathbb{N}$.

Thus, for all $k \in \mathbb{N}$ sufficiently large, $\lambda = 2^k$ and $s = -\epsilon$ as above, these estimates and the assumed embedding give

$$2^{-k(d/p+\epsilon/2)} \lesssim \|f^{\lambda}\|_{\mathcal{M}^{p,q}} \lesssim \|f^{\lambda}\|_{B_s^{p,q}} \lesssim 2^{-k(d/p+\epsilon)}.$$

This is clearly a contradiction, hence $s \geq 0 = d\theta_1(p, q)$. The case $(p, q) = (1, \infty)$ is obtained as follows. Pick $\varphi \in \mathscr{C}_c^{\infty}$ with supp $(\widehat{\varphi}) \subset \{\xi : 1/2 \leq \xi \leq 2\}$. By our assumption, $B_s^{1,\infty} \subset \mathcal{M}^{1,\infty} \subset \mathscr{F}L^{\infty}$. Now, since for all $k \in \mathbb{N}$ we have (with $\lambda = 2^k$)

$$2^{-kd} \lesssim \|\varphi^{\lambda}\|_{\mathcal{M}^{1,\infty}} \lesssim \|\varphi^{\lambda}\|_{B_s^{1,\infty}} \lesssim 2^{k(s-d)},$$

we obtain $s \geq 0$. $\qquad\qquad\qquad\qquad\qquad\qquad\qquad\qquad\qquad\qquad\qquad\qquad\square$

The other cases follow essentially from similar considerations. The interested reader is referred to [211] for further details. We end this section by stating the optimal embedding result between weighted modulation spaces and Besov spaces, which generalizes Theorems 6.3 and 6.6; see for example [204] and the references therein.

Theorem 6.7 *Let $0 < p, q \leq \infty$ and $s_1, s_2 \in \mathbb{R}$. The following are true:*

(1) $B_{s_1}^{p,q}(\mathbb{R}^d) \subset \mathcal{M}_{s_2}^{p,q}(\mathbb{R}^d)$ *if and only if* $s_1 - s_2 \geq d\theta_1(p, q)$.
(2) $\mathcal{M}_{s_2}^{p,q}(\mathbb{R}^d) \subset B_{s_1}^{p,q}(\mathbb{R}^d)$ *if and only if* $s_1 - s_2 \leq d\theta_2(p, q)$.

An immediate consequence of Theorem 6.7 is that we also have the following embeddings:

(i) $B_{s+d/2}^{2,1}(\mathbb{R}^d) \subset \mathcal{M}_s^{2,1}(\mathbb{R}^d) \subset B_s^{2,1}(\mathbb{R}^d)$.
(ii) $B_{s+d}^{\infty,1}(\mathbb{R}^d) \subset \mathcal{M}_s^{\infty,1}(\mathbb{R}^d) \subset B_s^{\infty,1}(\mathbb{R}^d)$.

6.2 Embeddings Between Modulation Spaces and L^p-Sobolev Spaces

We have the following embedding result.

Theorem 6.8 *Let* $1 \leq p, q \leq \infty$ *and* $s \in \mathbb{R}$.

(a) *If* $s > d\theta_1(p, q)$, *then* $L_s^p(\mathbb{R}^d) \subset \mathcal{M}^{p,q}(\mathbb{R}^d)$. *Conversely, if* $L_s^p(\mathbb{R}^d) \subset \mathcal{M}^{p,q}(\mathbb{R}^d)$, *then* $s \geq d\theta_1(p, q)$.
(b) *If* $s < d\theta_2(p, q)$, *then* $\mathcal{M}^{p,q}(\mathbb{R}^d) \subset L_s^p(\mathbb{R}^d)$. *Conversely, if* $\mathcal{M}^{p,q}(\mathbb{R}^d) \subset L_s^p(\mathbb{R}^d)$, *then* $s \leq d\theta_2(p, q)$.

Proof This follows immediately from Theorems 6.3 and 6.6, together with the well-known embedding [229, p. 97]

$$L_{s+\epsilon}^p(\mathbb{R}^d) \subset B_s^{p,q}(\mathbb{R}^d) \subset L_{s-\epsilon}^p(\mathbb{R}^d), \ \epsilon > 0.$$

$\square$

The slight drawback in the statement of the anterior theorem has to do with the sufficiency for inclusion in the "critical" cases $s = d\theta_1(p, q)$ and $s = d\theta_2(p, q)$. The complete answer about optimality was provided by Kobayashi and Sugimoto [170, Theorem 1.3, Theorem 1.4].

Theorem 6.9 *Let* $1 \leq p, q \leq \infty$ *and* $s \in \mathbb{R}$. *Then* $L_s^p(\mathbb{R}^d) \subset \mathcal{M}^{p,q}(\mathbb{R}^d)$ *if and only if one of the following conditions is satisfied:*

(1) $q \geq p > 1$ *and* $s \geq d\theta_1(p, q)$;
(2) $p > q$ *and* $s > d\theta_1(p, q)$;
(3) $p = 1, q = \infty$, *and* $s \geq d\theta_1(1, \infty)$;
(4) $p = 1, q \neq \infty$ *and* $s > d\theta_1(1, q)$.

Theorem 6.10 *Let* $1 \leq p, q \leq \infty$ *and* $s \in \mathbb{R}$. *Then* $\mathcal{M}^{p,q}(\mathbb{R}^d) \subset L_s^p(\mathbb{R}^d)$ *if and only if one of the following conditions is satisfied:*

(1) $q \leq p < \infty$ *and* $s \leq d\theta_2(p, q)$;
(2) $p < q$ *and* $s < d\theta_2(p, q)$;
(3) $p = \infty, q = 1$, *and* $s \leq d\theta_2(\infty, 1)$;
(4) $p = \infty, q \neq 1$, *and* $s < d\theta_2(\infty, q)$.

The proofs of Theorems 6.9 and 6.10 are in turn based on arguments of Kobayashi et al. [171], the embedding between weighted modulation spaces $\mathcal{M}_s^{p,q}$ and local Hardy spaces on one hand and L^p-Sobolev spaces on the other hand.

6.3   Notes

Theorems 4.42 and 4.43 stated the boundedness of the multiplier operator $L_{\sigma_{t;\alpha}}$ from $\mathscr{M}^{p,q}(\mathbb{R}^d)$ to $\mathscr{M}^{p,q}(\mathbb{R}^d)$ for $\alpha \in [0,2]$ and from $\mathscr{M}_s^{p,q}(\mathbb{R}^d)$ to $\mathscr{M}^{p,q}(\mathbb{R}^d)$ for $\alpha > 2$ if and only if s is larger than an appropriate index $s_0 = s_0(\alpha, d, p)$. A similar result due to Miyachi [182] gives the boundedness of $L_{\sigma_{t;\alpha}}$ from $L_s^p(\mathbb{R}^d)$ to $L^p(\mathbb{R}^d)$ for $\alpha > 1$ if and only if s is larger than $s_1 = s_0 + 2d|1/p - 1/2|$; in particular, this quantifies precisely the loss in regularity needed to compensate for the fact that, in general, $L_{\sigma_{t;\alpha}}$ is not bounded on $L^p(\mathbb{R}^d)$. The embeddings between L^p-Sobolev spaces and modulation spaces can be used to characterize precisely the boundedness of $L_{\sigma_{t;\alpha}}$ from $\mathscr{M}_s^{p,q}$ to L^p and from L_s^p to $\mathscr{M}^{p,q}$ in terms of the relationships between the indices p, q, and s. For example, the following theorems hold, see [171, Corollary 5.2, Corollary 5.4].

Theorem 6.11 *Let* $1 \leq p,q \leq \infty$, $s,t \in \mathbb{R}$, *and* $0 \leq \alpha \leq 2$. *Then* $L_{\sigma_{t;\alpha}}$ *is bounded from* $L_s^p(\mathbb{R}^d)$ *to* $\mathscr{M}^{p,q}(\mathbb{R}^d)$ *if and only if one of the following conditions is satisfied:*

(1) $q \geq p > 1$ *and* $s \geq d\theta_1(p,q)$;
(2) $p > q$ *and* $s > d\theta_1(p,q)$;
(3) $p = 1, q = \infty$, *and* $s \geq d\theta_1(1,\infty)$;
(4) $p = 1, q \neq \infty$, *and* $s > d\theta_1(1,q)$.

Theorem 6.12 *Let* $1 \leq p,q \leq \infty$, $s,t \in \mathbb{R}$, *and* $\alpha > 2$. *Then* $L_{\sigma_{t;\alpha}}$ *is bounded from* $L_s^p(\mathbb{R}^d)$ *to* $\mathscr{M}^{p,q}(\mathbb{R}^d)$ *if one of the following conditions is satisfied:*

(1) $q \geq p > 1$ *and* $s \geq d\theta_1(p,q) + s_0$;
(2) $p > q$ *and* $s > d\theta_1(p,q) + s_0$;
(3) $p = 1, q = \infty$, *and* $s \geq d\theta_1(1,\infty) + s_0$;
(4) $p = 1, q \neq \infty$, *and* $s > d\theta_1(1,q) + s_0$.

A partial converse of Theorem 6.12 also holds, see [171, Theorem 5.5].

Theorem 6.13 *Let* $1 \leq p,q \leq \infty$, $s,t \in \mathbb{R}$, *and* $\alpha > 2$. *If* $L_{\sigma_{t;\alpha}}$ *is bounded from* $L_s^p(\mathbb{R}^d)$ *to* $\mathscr{M}^{p,q}(\mathbb{R}^d)$, *then* $s \geq d\theta_1(p,q) + s_0$.

For the boundedness of $L_{\sigma_{t;\alpha}}$ from $\mathscr{M}_s^{p,q}(\mathbb{R}^d)$ to $L^p(\mathbb{R}^d)$ one simply needs to write the dual statements to the ones above.

Chapter 7
Applications to Partial Differential Equations

7.1 The Schrödinger Equation

The discovery by Schrödinger in the early 1920s of the linear partial differential equation that bears his name as being the one that governs the state of a quantum system in terms of the wave function, of position and time, constitutes one of the fundamental results of quantum mechanics. Its nonlinear versions are equally important as models of waves phenomena in nonlinear optics and plasma physics and they have received lots of attention from a theoretical point of view as well, spurring some exciting developments of new ideas and analytical tools in linear and multilinear harmonic analysis. Both the linear and nonlinear Schrödinger equations are examples of *dispersive partial differential equations*, expressing the fact that plane-waves with different wavelengths will travel at different velocities. To explain in simple terms how dispersion works [215], let us take the example of the so-called Airy equation, which is an example of a linear wave equation:

$$\partial_t u + \partial_x^3 u = 0, \ (x, t) \in \mathbb{R} \times \mathbb{R}. \tag{7.1}$$

A solution of the form $u_{k,\omega}(x, t) = e^{i(kx-\omega t)}$ is called a *plane-wave solution*. The parameters k and ω are referred to as the wave number and frequency, respectively, while the velocity of the traveling wave is given by the ratio ω/k. By substituting the expression of such a plane-wave solution in (7.1), we obtain the *dispersion relation* $\omega/k = -k^2$, effectively determining a unique frequency $\omega(k)$ associated with a wave number k. Furthermore, as the dispersion relation makes it clear, the higher the frequency (wave number), the faster the plane-wave travels. By superposition, if one would prescribe an initial profile $u(x, 0) = f(x)$, the solution given by

$$u(x, t) = \int \widehat{f}(k) e^{i(kx-\omega(k)t)} \, dk$$

© Springer Science+Business Media, LLC, part of Springer Nature 2020
Á. Bényi, K. A. Okoudjou, *Modulation Spaces*, Applied and Numerical
Harmonic Analysis, https://doi.org/10.1007/978-1-0716-0332-1_7

would display the same behavior of different Fourier modes traveling at different velocities; for example, an initial Gaussian profile $f(x) = e^{-x^2}$ would flatten out over time.

From both an applied and theoretical point of view, questions related to the existence, uniqueness, and stability of solutions to nonlinear dispersive partial differential equations, their long-time behavior and the formation of singularities all play central roles. These topics are encompassed under the name of (local or global) well posedness theory. In the classical setting of the well-posedness theory, the main goal is the existence and uniqueness of solutions for all initial profiles in Sobolev spaces $H^s(\mathbb{R}^d)$, for appropriate regularity indices s dictated by a critical threshold regularity dependent on the dimension d and the nonlinearity. Without being entirely rigorous, a generic principle governing well posedness is that when the dispersion dominates the concentration introduced by the nonlinearity, we expect well posedness. Lowering the index s beyond the critical regularity creates pathological behavior, and in such case one may be content with showing that almost all solutions behave well, thus yielding a probabilistic well-posedness theory, an exciting interdisciplinary area of study becoming increasingly popular. In this context, the intuition provided by the simple time-frequency decomposition defining the modulation spaces becomes extremely useful; see the Notes at the end of this chapter.

Another approach is to study the question of well posedness on Fourier-Lebesgue spaces $\mathscr{F}L^{s,p}$. For the modified Kortweg-de Vries equation on the real line such questions naturally lead to the modulation spaces via the embedding $\mathscr{F}L^{s,p}(\mathbb{R}) \subset \mathscr{M}_s^{2,p}(\mathbb{R})$, $p \geq 2$, and the calculations in this case become much more involved [191]; see also [43]. In general, while the well-posedness results may not be directly comparable to the ones from the classical deterministic theory on Sobolev spaces, the fact remains that the modulation spaces are *low regularity spaces* and, as such, understanding the solutions with appropriate rough initial profiles becomes important. For example, in the case of the NLS (7.7) we can prove its well posedness in $\mathscr{M}_s^{2,1}(\mathbb{R}^d)$ for all odd exponents $k \in 2\mathbb{N} + 1$ and this space contains profiles that cannot be handled in the super-critical Sobolev regime; see Theorem 7.4 and also [140] for an extension to the setting of α-modulation spaces.

In this chapter, we will show that the linear and nonlinear Schrödinger equations behave well with respect to modulation spaces. The essential insight is provided by the automatic decomposition on high-low frequencies and the "naturally" available nonlinear estimates on these spaces. We also indicate the interactions between the Galilean transformation and the Fourier and short-time Fourier transforms. More generally, this chapter illustrates the potential role the modulation spaces can play in the analysis of nonlinear dispersive partial differential equations. It is a fitting conclusion to the monograph and, we hope, the beginning and the continuation of the still fresh role of these spaces in PDEs.

7.1.1 The Linear Schrödinger Equation

For $(x, t) \in \mathbb{R}^d \times \mathbb{R}$, $u(x, t)$ satisfies the (linear) *Schrödinger equation* if

$$i \partial_t u + \Delta u = 0. \tag{7.2}$$

For the corresponding Cauchy problem, we want to find the solution u for which we are also prescribed an initial data, $u(x, 0) = f(x)$. If we fix the time variable t and formally take the Fourier transform in the space variable x, we see that

$$u(x, t) = S(t)(f)(x) := e^{it\Delta}(f)(x),$$

where $S(t)$ is the *Fourier multiplier* operator given by

$$\widehat{S(t)(f)}(\xi) = e^{-i4\pi^2 t |\xi|^2} \widehat{f}(\xi). \tag{7.3}$$

If we invert again formally, we get, with the notation of Theorem 4.42,

$$u(x, t) = (\widehat{S(t)(f)})^\vee(x) = L_{\sigma_{-4\pi^2 t}}(f)(x) = \int_{\mathbb{R}^d} e^{2\pi i x \cdot \xi} e^{-i4\pi^2 t |\xi|^2} \widehat{f}(\xi) \, d\xi.$$

These manipulations can obviously be justified if the initial data is sufficiently smooth. For example, if we assume that $f \in \mathscr{S}$, then $S(t)(f) \in \mathscr{S}$ for all t and $u(x, t) = S(t)(f)(x)$ is a $\mathscr{C}^\infty$ function that solves the Cauchy problem of (7.2) with initial data f. Because of the density of $\mathscr{S}$ in L^2, the last assertion can be extended to the setting of square integrable initial data. More precisely, if $f \in L^2(\mathbb{R}^d)$, $S(t)(f)(x)$ solves (7.2) in the sense of distributions; that is, for any test function $\phi \in \mathscr{C}_c^\infty(\mathbb{R}^d \times \mathbb{R})$, we have

$$\int_{\mathbb{R}^d} (i\partial/\partial t + \Delta)(S(t)(f))(x)\phi(x, t) \, dxdt = \int_{\mathbb{R}^d} (i\phi_t - \Delta\phi)(x)S(t)(f)(x) \, dxdt = 0.$$

Viewed on the spatial side, $S(t)$ is an operator of convolution type,

$$S(t)(f) = K_t * f, \quad \text{where} \quad K_t(x) = (4\pi i t)^{-d/2} e^{-|x|^2/(4it)}.$$

Thus, Plancherel's theorem and Young's inequality immediately give that

$$\|S(t)(f)\|_{L^2} = \|f\|_{L^2} \quad \text{and} \quad \|S(t)(f)\|_{L^\infty} \lesssim |t|^{-d/2} \|f\|_{L^1}.$$

In fact, a careful application of the Hausdorff-Young inequality gives the *decay estimate*

$$\|S(t)(f)\|_{L^{p'}} \lesssim |t|^{-d(1/p - 1/2)} \|f\|_{L^p}, \quad 1 \le p \le 2.$$

It is natural then to ask what other $L^p \to L^r$ estimates, with $r < \infty$, could one expect for the multiplier $S(t)$. An important notion in this context is that of *scaling invariance*; it is easy to show that if u solves the Cauchy problem of (7.2) with initial data f, then, for all $a > 0$, $u_a(x,t) = u(ax, a^2 t)$ solves the Cauchy problem with initial data $f_a(x) = f(ax)$. Note now that if we have

$$\|u\|_{L^r(\mathbb{R}^d \times \mathbb{R})} = \|S(t)(f)\|_{L^r(\mathbb{R}^d \times \mathbb{R})} \lesssim \|f\|_{L^p(\mathbb{R}^d)}, \tag{7.4}$$

then necessarily we also have

$$\|u_a\|_{L^r(\mathbb{R}^d \times \mathbb{R})} = \|S(t)(f_a)\|_{L^r(\mathbb{R}^d \times \mathbb{R})} \lesssim \|f_a\|_{L^p(\mathbb{R}^d)}.$$

But since $\|u_a\|_{L^r} = a^{-(d+2)/r}\|u\|_{L^r}$ and $\|f_a\|_{L^p} = a^{-d/p}\|f\|_{L^p}$, we must have

$$\frac{2}{r} + \frac{d}{r} = \frac{d}{p} \Leftrightarrow r = \frac{d+2}{d}p.$$

For a proof of the *Strichartz estimate* (7.4) when $p = 2$, see for example [209, Theorem 6.3]. The estimate (7.4) can be generalized to include all so-called admissible pairs, see [118, 165, 210, 232]. In our setting, a pair (q, r) is said to be *Schrödinger admissible* if it satisfies

$$\frac{2}{q} + \frac{d}{r} = \frac{d}{2} \tag{7.5}$$

with $2 \le q, r \le \infty$ and $(q, r, d) \ne (2, \infty, 2)$. We have the following *homogeneous Strichartz estimate*.

Proposition 7.1 *If (q, r) is Schrödinger admissible, then*

$$\|S(t)(f)\|_{L^r_x L^q_t(\mathbb{R}^d \times \mathbb{R})} \lesssim \|f\|_{L^2(\mathbb{R}^d)}. \tag{7.6}$$

It is worth pointing out that such Strichartz estimates also exist in the context of Wiener amalgam spaces, but they are incompatible with (7.6) in the sense that they cannot be deduced from (7.6) nor do they imply it; see [64, 65], and also [13].

7.1.2 The Nonlinear Schrödinger Equation

Of interest to us will be the *nonlinear Schrödinger equation* (NLS) with power-nonlinearity of the form

$$i\partial_t u + \Delta u = \pm|u|^{k-1}u. \tag{7.7}$$

Here, the exponent $k > 1$. Assume, as in the previous subsection, that the initial condition $u(x, 0) = f(x)$ is in $L^p(\mathbb{R}^d)$. We modify slightly the ideas exposed there and require that if u solves the NLS Cauchy problem with initial data f, then $u_{a,\gamma}(x, t) = a^\gamma u(ax, a^2 t)$ solves the NLS Cauchy problem with initial data $f_{a,\gamma}(x) = a^\gamma f(ax)$ for all $a > 0$. In order to conserve the L^p norm of the initial data under these scalings, we must have $\gamma = d/p$, while for $u_{a,\gamma}$ to solve (7.7) we necessarily have $\gamma + 2 = k\gamma$, thus $k = (\gamma + 2)/\gamma = (d + 2p)/d$.

Let us write for simplicity $N = N(u)$ for the nonlinearity $\pm |u|^{k-1} u$. Then, the Cauchy problem corresponding to the NLS with initial data f can be written as

$$u(x, t) = S(t)(f)(x) - i\mathscr{I}(N)(x, t), \tag{7.8}$$

where

$$\mathscr{I}(N)(x, t) = \int_0^t S(t - s) N(x, s)\, ds.$$

Here, we understand that, for each t and s, the operator $S(t - s)$ acts on $N(x, s)$ as a function of x. Using again a scaling argument, we see that in order for

$$\|\mathscr{I}(N)\|_{L^r(\mathbb{R}^d \times \mathbb{R})} \lesssim \|N\|_{L^s(\mathbb{R}^d \times \mathbb{R})}$$

to be true we must now have $s = r/k$; here, $r = \frac{d+2}{d} p$ is the exponent found in the previous subsection, for which $S(t)f \in L^r$ when the initial condition $f \in L^p$. In general, we have the following *inhomogeneous Strichartz estimate*.

Proposition 7.2 *If (q, r) is Schrödinger admissible, then*

$$\|\mathscr{I}(N)\|_{L_x^r L_t^q (\mathbb{R}^d \times \mathbb{R})} \lesssim \|N\|_{L_x^{r'} L_t^{q'} (\mathbb{R}^d \times \mathbb{R})}. \tag{7.9}$$

Ultimately, the (homogeneous and inhomogeneous) Strichartz estimates are tools used in the search for a solution to the NLS Cauchy problem. By definition, such a solution would be a function $u \in L^r(\mathbb{R}^d \times \mathbb{R})$ such that for all $f \in L^p$, u satisfies (7.7) in the sense of distributions and, for each $t \in \mathbb{R}$, $u(\cdot, t) \in L^p(\mathbb{R}^d)$, the mapping $\mathbb{R} \ni t \to u(\cdot, t) \in L^p(\mathbb{R}^d)$ is continuous, and $u(x, 0) = f(x)$ for all $x \in \mathbb{R}^d$. That such a solution exists turns out to be correct if we require the L^p norm of the initial condition f to be sufficiently small. For a proof of this fact when $p = 2$, see [209, Theorem 6.9]; the strong solution is obtained as a fixed point of an "obvious" contraction mapping defined on a closed ball in $L^r(\mathbb{R}^d \times \mathbb{R})$ of suitably small radius. We will revisit this fixed-point argument in our discussion of well posedness of the NLS on modulation spaces.

7.1.3 The Galilean Transformation

We have seen already that given $\alpha > 0$ and a solution u of (7.7), we can construct a dilated family of new solutions $u_\alpha(x, t) = \alpha^{d/2} u(\alpha x, \alpha^2 t)$ for $k = (d + 4)/d$ and $p = 2$ in (7.7). We wish to show now that there is another way of constructing new families of solutions to (7.7) that is custom made for time-frequency analysis.

Given a solution $u(x, t)$ of (7.7), we can obtain a family of new solutions $\{u^a\}_{a \in \mathbb{R}^d}$ via the *Galilean transformation*:

$$u^a(x, t) = e^{i\frac{a}{2} \cdot x} e^{i\frac{|a|^2}{4} t} u(x + at, t). \tag{7.10}$$

In other words, the NLS is invariant under the Galilean transformation.

Let us fix the time variable t and take the Fourier transform of u^a with respect to the space variable. We have

$$\widehat{u^a}(\xi, t) = \left(e^{i\frac{a}{2} \cdot x} e^{i\frac{|a|^2}{4} t} u(\cdot + at, t)\right)^\wedge(\xi) = e^{i\frac{|a|^2}{4} t} \left(u(\cdot + at, t)\right)^\wedge(\xi - \tfrac{a}{4\pi})$$

$$= e^{i\frac{|a|^2}{4} t} e^{2\pi i at \cdot \xi} e^{-i\frac{|a|^2}{2} t} \widehat{u}(\xi - \tfrac{a}{4\pi}, t) = e^{-i\frac{|a|^2}{4} t} e^{2\pi i at \cdot \xi} \widehat{u}(\xi - \tfrac{a}{4\pi}, t).$$

In particular, we obtain that $|\widehat{u^a}(\xi, t)| = |\widehat{u}(\xi - \tfrac{a}{4\pi}, t)|$.

Similarly, note that for a fixed t, we can write

$$u^a(x, t) = e^{i\frac{|a|^2}{4} t} M_{\frac{a}{4\pi}} T_{-at} u(\cdot, t)(x);$$

we recall that the notations M_x and T_y used here stand for the modulation (by x) and translation (by y) operators defined in Sect. 1.2.4. By applying now the STFT (with some fixed window ϕ) and using the almost commutation property of time-frequency shifts (1.9) and the covariance property of the STFT (see Proposition 1.82), we have

$$V_\phi u^a(x, \xi) = e^{i\frac{|a|^2}{4} t} V_\phi(e^{-i\frac{|a|^2}{2} t} T_{-at} M_{\frac{a}{4\pi}} u(\cdot, t))(x, \xi)$$

$$= e^{-i\frac{|a|^2}{4} t} e^{2\pi i at \cdot \xi} V_\phi u\left(x + at, \xi - \frac{a}{4\pi}\right).$$

In particular, $|V_\phi u^a(x, \xi)| = \left|V_\phi u\left(x + at, \xi - \frac{a}{4\pi}\right)\right|$.

The previous two calculations show that the Galilean transformation preserves the $\mathscr{M}^{p,q}$ norm for all $a \in 2\mathbb{Z}^d$. Furthermore, for fixed t, we have the norm equivalence $\|u^a\|_{\mathscr{M}^{p,q}} \sim \|u\|_{\mathscr{M}^{p,q}}$ for all $a \in \mathbb{R}^d$, where the implicit constant is independent of a. Note, however, that due to the additional shift on the Fourier side, the Galilean transformation *does not* preserve the weighted modulation space $\mathscr{M}^{p,q}_s$ norm for $s \neq 0$.

The invariance of the NLS under the Galilean transformation (which is defined in terms of time-frequency shifts) is a strong hint that this equation should be well behaved when studied on modulation spaces. The following section is devoted to the study of the Schrödinger equation on these spaces.

7.2 Well Posedness of NLS on Modulation Spaces

The $\pm$ sign of the coefficient in the algebraic nonlinearity $|u|^{k-1}u$ from (7.7) determines the focusing or defocusing character of the equation. This character is of relevance when studying the well posedness of NLS on, say, Sobolev spaces. Similarly, the scaling symmetry and the dispersive effect of the associated linear flow $S(t)$ all play important roles; we note that the Strichartz estimates are often just an expression of dispersion. For an introduction to these and other related topics, the interested reader can consult [216].

Before we continue with the concept of well posedness, we point out that the NLS has also a *time reversal symmetry*; that is, if $u(x, t)$ solves (7.7), so does $\bar{u}(-t, x)$. Because of this symmetry, the behavior of a solution for $t > 0$ is similar to the one for $t < 0$; as such, we can restrict to the forward-in-time direction in our discussion of well posedness.

Roughly speaking, well posedness amounts to existence, uniqueness, and continuous dependence on initial data of solutions. Let $(\mathscr{X}, \| \cdot \|)$ be a normed space of functions on $\mathbb{R}^d$. We say that (7.7) is *locally well posed in* $\mathscr{X}$, if for a given initial data $f \in \mathscr{X}$, there exists a $\tau = \tau(\|f\|_{\mathscr{X}}) > 0$ and a unique $u \in \mathscr{C}([0, \tau]; \mathscr{X})$ that solves (7.7) and the solution map

$$V \ni f \mapsto u \in \mathscr{C}([0, \tau]; \mathscr{X})$$

is continuous; here, we wrote $f(x) = u(x, 0)$ for the initial data. Writing $u(t)(x) = u(x, t)$, we can view $u(t) : \mathbb{R}^d \to \mathbb{C}$ as a function on $\mathbb{R}^d$ and the notation $u \in \mathscr{C}([0, \tau]; \mathscr{X})$ means that $u(t) \in \mathscr{X}$ for all $t \in [0, \tau]$, with $[0, \tau] \ni t \mapsto u(t) \in \mathscr{X}$ continuous; in particular, $[0, \tau] \ni t \to \|u(t)\|_{\mathscr{X}} \in [0, \infty)$ is continuous. The natural norm induced on the space $\mathscr{C}([0, \tau]; \mathscr{X})$ is given by

$$\|u\| = \sup_{t \in [0, \tau]} \|u(t)\|_{\mathscr{X}}.$$

Similarly, we say that (7.7) is *globally well posed in* $\mathscr{X}$, if *the time of existence* τ in the local well posedness can be made arbitrarily large.

In what follows, consider the generic NLS

$$i\partial_t u + \Delta u = N(u),$$

with the nonlinear term $N(u)$ not necessarily power-like as in (7.7). The problem of existence (and uniqueness) of $u \in \mathscr{C}([0, \tau]; \mathscr{X})$ that solves the corresponding Cauchy problem with initial data $u(x, 0) = f(x)$ will be, by definition, the same as for the *Duhamel formulation* of the problem:

$$u(t) = S(t)(f) - i \int_0^t S(t - t')N(u(t')) \, dt'. \tag{7.11}$$

Thus, further denoting by

$$T(u)(t) = S(t)(f) - i \int_0^t S(t - t')N(u(t')) \, dt', \tag{7.12}$$

we are in effect looking for *a fixed point u in some space $\mathscr{X}$ for the mapping T*; that is, $u \in \mathscr{X}$ such that $T(u) = u$. As such, *Banach's fixed point theorem (or contraction mapping principle)* plays an important role.

Theorem 7.3 *Let $(\mathscr{X}, \| \cdot \|)$ be a Banach space. Assume that $T : \mathscr{X} \to \mathscr{X}$ is a contraction mapping on $\mathscr{X}$, that is, there exists a constant $\lambda \in [0, 1)$ such that for all $u, v \in \mathscr{X}$,*

$$\|T(u) - T(v)\| \le \lambda \|u - v\|.$$

Then, T has a unique fixed point in $\mathscr{X}$.

A proof of this classical result can be found in Appendix A. Returning to the Duhamel principle, we note that this idea is quite natural if one recalls the Picard-Lindelöf or Peano theorems that establish the existence and uniqueness of solutions to any ODE for a continuous field. For the statements and sketches of proofs of these results, see Appendix B.

Our first result about local well-posedness concerns (7.7) in which the power k is a positive odd integer. In the following, the sign $\pm$ plays no role in the discussion and will be ignored.

Theorem 7.4 *Let $k \in \mathbb{N} \cup \{0\}$ and $N(u) = |u|^{2k}u$. Then, the NLS (7.7) is locally well posed in $\mathscr{M}_s^{p,1}$ for all $p \ge 1$ and $s \ge 0$.*

Proof Let $u(\cdot, 0) = f$ be in $\mathscr{M}_s^{p,1}$. We are trying to show that there exists $u \in \mathscr{C}([0, \tau]; \mathscr{M}_s^{p,1})$, where $0 < \tau < \infty$ (to be chosen later), satisfying (7.11). Let $t \in [0, \tau]$. By essentially repeating the calculations in Proposition 4.41, we find that

$$\|S(t)(f)\|_{\mathscr{M}_s^{p,1}} \lesssim (1 + t^2)^{d/4} \|f\|_{\mathscr{M}_s^{p,1}} \lesssim (1 + \tau^2)^{d/4} \|f\|_{\mathscr{M}_s^{p,1}}. \tag{7.13}$$

Moreover, using (7.13) and (5.3), we have

$$\left\| \int_0^t S(t - t') N(u(t')) \, dt' \right\|_{\mathscr{C}([0,\tau]; \mathscr{M}_s^{p,1})} \leq \int_0^t \| S(t - t') N(u(t')) \|_{\mathscr{C}([0,\tau]; \mathscr{M}_s^{p,1})} \, dt'$$

$$\lesssim \tau (1 + \tau^2)^{d/4} \| u \|_{\mathscr{C}([0,\tau]; \mathscr{M}_s^{p,1})}^{2k+1}.$$

(7.14)

Combining (7.13) and (7.14), we obtain the following estimate on the operator T defined in (7.12):

$$\| T(u) \|_{\mathscr{C}([0,\tau]; \mathscr{M}_s^{p,1})} \leq c(1 + \tau^2)^{d/4} (\| f \|_{\mathscr{M}_s^{p,1}} + \tau \| u \|_{\mathscr{C}([0,\tau]; \mathscr{M}_s^{p,1})}^{2k+1}), \qquad (7.15)$$

for some universal constant $c > 0$.

Let now $r = 2c(1 + \tau^2) \| f \|_{\mathscr{M}_s^{p,1}}$ and $\mathscr{X}$ denote the closed ball centered at the origin and of radius r in $\mathscr{C}([0, \tau]; \mathscr{M}_s^{p,1})$, that is

$$\mathscr{X} := \{ u \in \mathscr{C}([0, \tau]; \mathscr{M}_s^{p,1}) : \| u \| \leq r \};$$

here, $\| \cdot \|$ denotes the induced norm of $\mathscr{C}([0, \tau]; \mathscr{M}_s^{p,1})$ on $\mathscr{X}$. Further, consider τ so that $c\tau(1 + \tau^2)^{d/4} r^{2k} \leq 1/2$; this is clearly equivalent to saying that $\tau < \widetilde{\tau}$, where $\widetilde{\tau} = \widetilde{\tau}(\| f \|_{\mathscr{M}_s^{p,1}})$ is dependent on the modulation space norm of the initial data f. Then, if $u \in \mathscr{X}$, we have

$$\| T(u) \| \leq \frac{r}{2} + \frac{r}{2} = r,$$

that is, $T(u) \in \mathscr{X}$ as well. We have just proved that T is a mapping from $\mathscr{X}$ to $\mathscr{X}$.

To apply Banach's fixed point theorem, it remains to show that T is a contraction on $\mathscr{X}$. In what follows, $u, v \in \mathscr{X}$. First, we point out the following simple estimate:

$$\| |u|^2 - |v|^2 \| = \| \bar{u}(u - v) + v(\overline{u - v}) \| \leq 2r \| u - v \|.$$

Noting now that

$$N(u) - N(v) = |u|^{2k}(u - v) + v(|u|^{2k} - |v|^{2k})$$

and

$$|u|^{2k} - |v|^{2k} = (|u|^2 - |v|^2) \sum_{j=1}^k |u|^{2k-2j} |v|^{2j-2}$$

we obtain

$$\| N(u) - N(v) \| \leq (2k + 1) r^{2k} \| u - v \|.$$

Therefore, selecting $\tilde{\tau}$ appropriately smaller so that $2(2k+1)\tau(1+\tau)^{d/4}r^{2k} \leq 1$ for $\tau < \tilde{\tau}$, we obtain that

$$\|T(u) - T(v)\| \leq \frac{1}{2}\|u - v\|.$$

We have shown that T is a contraction on $\mathscr{X}$. Thus, T has a fixed point in $\mathscr{X}$, which is our local solution (on $[0, \tilde{\tau}]$) to (7.7). For the continuity of the solution map, let $\tilde{u}$ be a solution of (7.7) corresponding to the initial condition $\tilde{f}$. Using again (7.11), we have

$$\|u - \tilde{u}\|_{\mathscr{C}([0,\tau];\mathscr{M}_s^{p,1})} \leq c(1+\tau^2)^{d/4}(\|f - \tilde{f}\|_{\mathscr{M}_s^{p,1}} + \tau\|N(u) - N(\tilde{u})\|_{\mathscr{C}([0,\tau];\mathscr{M}_s^{p,1})}).$$

So, for $0 < \tau < \tilde{\tau}$ sufficiently small and $u, \tilde{u} \in \mathscr{X}$, we have

$$\|u - \tilde{u}\|_{\mathscr{C}([0,\tau];\mathscr{M}_s^{p,1})} \leq C\|f - \tilde{f}\|_{\mathscr{M}_s^{p,1}},$$

for some $C > 0$. This proves that, in fact the solution map is Lipschitz continuous and, in particular, locally uniformly continuous.

We note also that given $u, \tilde{u} \in \mathscr{C}_\tau := \mathscr{C}([0, \tau]; \mathscr{M}_s^{p,1})$ solutions to the NLS with $u(0) = f \in \mathscr{M}_s^{p,1}$, $\tilde{u}(0) = \tilde{f} \in \mathscr{M}_s^{p,1}$, a similar calculation to the one above and Gronwall's lemma (see Appendix C) applied to $\|u - \tilde{u}\|_{\mathscr{C}([0,t];\mathscr{M}_s^{p,1})}$ give

$$\|u - \tilde{u}\|_{\mathscr{C}([0,\tau];\mathscr{M}_s^{p,1})} \leq c_1(\tau)e^{c_2(\tau)(\|u\|_{\mathscr{C}_\tau}^{2k} + \|\tilde{u}\|_{\mathscr{C}_\tau}^{2k})}\|f - \tilde{f}\|_{\mathscr{M}_s^{p,1}},$$

showing the uniqueness of our solution u in the entire $\mathscr{C}_\tau$. This finishes the proof. □

We point out that the methodology used in the proof of Theorem 7.4 can be abstracted to produce a local well-posedness result on modulation spaces for any Duhamel formulation of the form

$$u(t) = V(t)(f) + \int_0^t V(t - t')L(u(t'))\,dt', \quad u(\cdot, 0) = f, \tag{7.16}$$

under suitable conditions on V and L. We state this precisely in the following result.

Theorem 7.5 *Let P, Q be two polynomials with positive coefficients and $h : [0, \infty) \to [0, \infty)$ be such that h is right-continuous at 0 and $h(0) = 0$. Let $p, q \in (0, \infty]$ and $s \in \mathbb{R}$. Assume that $\|V(t)(f)\|_{\mathscr{M}_s^{p,q}} \lesssim P(t)\|f\|_{\mathscr{M}_s^{p,q}}$, $\|L(u(t))\|_{\mathscr{M}_s^{p,q}} \lesssim Q(\|u(t)\|_{\mathscr{M}_s^{p,q}})$ and*

$$\|L(u) - L(v)\| \leq h(r)\|u - v\|$$

for all $\|u\|, \|v\| \leq r$. Then, (7.16) is locally well posed in $\mathscr{M}_s^{p,q}$.

Proof We simply indicate the essential parts of the argument, the details being left to the interested reader. Consider now the map

$$W(u)(t) = V(t)(f) + \int_0^t V(t - t')L(u(t'))\, dt',$$

with $t \in [0, \tau]$ and $\tau < \tilde{\tau} < \infty$ to be chosen shortly. The hypothesis implies the following bound on the operator W:

$$\|W(u)\|_{\mathscr{C}([0,\tau]; \mathscr{M}_s^{p,q})} \lesssim P(\tau)(\|f\|_{\mathscr{M}_s^{p,q}} + \tau\, Q(\|u\|_{\mathscr{C}([0,\tau]; \mathscr{M}_s^{p,q})})). \tag{7.17}$$

As in the proof of Theorem 7.4, we can now select a $\tilde{\tau} < \infty$ such that, for all $\tau < \tilde{\tau}$, we get $W : \mathscr{X} \to \mathscr{X}$; here, again,

$$\mathscr{X} := \{u \in \mathscr{C}([0, \tau]; \mathscr{M}_s^{p,q}) : \|u\| \leq r\},$$

with $r := r(\tilde{\tau})$ being so that $r(\tilde{\tau}) \to 0$ as long as $\tilde{\tau} \to 0^+$. Similarly, we obtain from the contraction like property of L that

$$\|W(u) - W(v)\| \leq c\tilde{\tau} P(\tilde{\tau})h(r(\tilde{\tau}))\|u - v\|.$$

Using now the right continuity of h at 0, we can select an appropriately smaller $\tilde{\tau}$ such that $2c\tilde{\tau} P(\tilde{\tau})h(r(\tilde{\tau})) \leq 1$, thus proving that W is a contraction on $\mathscr{X}$.

The continuity of the solution map follows just as in Theorem 7.4. □

Remark 7.6 It is tempting to infer, in view of Theorems 7.5 and 5.6 and its immediate corollary, that the local-well posedness of NLS on modulation spaces $\mathscr{M}_s^{p,2}$ is within reach for power-nonlinearities $N(u) = |u|^{k-1}u$ with k a non-integer. The difficulty, however, lies in surmounting the additional "real-valued" requirement from Theorem 5.6 and the fact that the functions operating on modulation spaces must have a special structure, see for example Theorem 5.9. Nevertheless, a more refined analysis, namely using the Fourier restriction method adapted to the modulation spaces, shows that for $d = 1$ and $p = 3$, the (cubic) NLS is locally well posed on all modulation spaces $\mathscr{M}_s^{2,p}$ with $2 \leq p < \infty$ and $s \geq 0$; see [134, Theorem 1.4] and [135]. Moreover, global well posedness can be obtained in the "almost critical" space $\mathscr{M}_s^{2,p}$; see [192]. The latter result is "almost critical" from the scaling point of view of the space $\mathscr{F}L^{0,\infty}$ which is embedded in $\mathscr{M}^{2,\infty}$. For a work on the derivative NLS which uses a similar strategy, the reader can consult [137]. For a notion of "critical exponent" in modulation spaces, see [49].

Returning to the NLS, we want to discuss briefly the issue of global well posedness. Generally speaking, the question about the global existence of a solution is a difficult one. For example, in the case of the cubic NLS (for which $N(u) = |u|^2 u$), the three known conservation laws,

$$\text{Mass: } M[u](t) := \int_{\mathbb{R}^d} |u(t, x)|^2 dx,$$

$$\text{Momentum: } P[u](t) := \text{Im} \int_{\mathbb{R}^d} u(t, x)\overline{\nabla u(t, x)}dx,$$

$$\text{Energy: } H[u](t) := \frac{1}{2} \int_{\mathbb{R}^d} |\nabla u(t, x)|^2 dx \pm \frac{1}{4} \int_{\mathbb{R}^d} |u(t, x)|^4 dx,$$

play an important role. In general, the existence of a local solution in the so-called critical Sobolev space usually leads to a global solution as long as the initial data is small; see [42]. In the context of modulation spaces, this generic principle permeates through. We illustrate it here by simply stating two global well-posedness results proved by Wang-Hudzik. The interested reader can consult [155] for the proofs and further discussions on this topic. In the next two theorems, $d, k \in \mathbb{N}$, the nonlinearity is $N = \pi_k$, where $\pi_k(u)$ denotes any k-product of u and $\bar{u}$, and k_0 is the positive root of the quadratic equation $dx^2 + (d-2)x - 4 = 0$. We also recall the notation P_k used for the smooth frequency-uniform decomposition operators, see Definition 3.13.

Theorem 7.7 *Let $k > k_0$ and $f \in \mathcal{M}^{1+1/k,1}$ be such that $\|f\|_{\mathcal{M}^{1+1/k,1}} \leq \delta$ for some $0 < \delta < 1$. Then, the NLS (7.11) has a unique global solution $u \in \mathcal{X}$, where*

$$\mathcal{X} := \{u \in \mathcal{S}'(\mathbb{R}^{d+1}) : \sup_{t \in \mathbb{R}} \langle t \rangle^{\frac{d(k-1)}{4(k+1)}} \|u(t)\|_{\mathcal{M}^{1+1/k,1}} < \infty\}.$$

It turns out that the initial data can be weakened further by only requiring it to belong to $\mathcal{M}^{2,1} \supset \mathcal{M}^{1+1/k,1}$. Due to the sharp embedding $B_{2,1}^{d/2} \subset \mathcal{M}^{2,1}$, the next result can be viewed as a global well-posedness result for small data in a "rougher" version of $B_{2,1}^{d/2}$.

Theorem 7.8 *Let $k \geq 1 + 4/d$ and $f \in \mathcal{M}^{2,1}$ be such that $\|f\|_{\mathcal{M}^{2,1}} \leq \delta$ for some $0 < \delta < 1$. Further assume that $q \in \mathbb{N}$ and $2 + 4/d \leq q \leq 1 + k$. Then, the NLS (7.11) has a unique global solution $u \in \mathcal{C}(\mathbb{R}; \mathcal{M}^{2,1}) \cap \mathcal{Y}_q$, where*

$$\mathcal{Y}_q := \{u \in \mathcal{S}'(\mathbb{R}^{d+1}) : (P_k u)_{k \in \mathbb{Z}} \in L_{x,t}^q \ell_{\mathbb{Z}}^1\}.$$

7.3 Notes

7.3.1. The well posedness result in Theorem 7.4 was for the NLS with an appropriate power nonlinearity. It is natural to consider the NLS with other nonlinearities $N(u)$ that are equally important in applications. For example, the NLS with exponential nonlinearity $N = e_\rho$, $e_\rho(u) = (e^{\rho|u|^2} - 1)u$, $\rho > 0$, is a model in optics. In fact, both the power nonlinearities of the form π_{2k+1} and the exponential-like ones e_ρ are particular cases of the ones that can be written as

$$N(u) = g(|u|^2)u, \tag{7.18}$$

where $g \in \mathscr{A}_+(\mathbb{C})$; here, we denote by $\mathscr{A}_+(\mathbb{C})$ the set of entire functions $g(z)$ with expansions of the form

$$g(z) = \sum_{k=1}^{\infty} c_k z^k, \, c_k \geq 0.$$

Observe that, in view of Theorem 5.9, the nonlinearities given by (7.18) are natural ones to consider. The following result is given in [14].

Theorem 7.9 *Let* $k \in \mathbb{N} \cup \{0\}$ *and* N *as in* (7.18). *Then, the NLS* (7.11) *is locally well posed in* $\mathscr{M}_s^{p,1}$ *for all* $p \geq 1$ *and* $s \geq 0$.

Moreover, the global results for power nonlinearities have counterparts in this realm as well. For example, we have the following statement paralleling that of Theorem 7.8, see [155].

Theorem 7.10 *Let* $k \geq 2$, $N = e_\rho$, *and* $f \in \mathscr{M}^{2,1}$ *be such that* $\|f\|_{\mathscr{M}^{2,1}} \leq \delta$ *for some* $0 < \delta < 1$. *Then, the NLS* (7.11) *has a unique global solution* $u \in \mathscr{C}(\mathbb{R}; \mathscr{M}^{2,1}) \cap \mathscr{Y}_4$.

As expected, the results presented for the NLS have counterparts for other nonlinear PDEs. Closely related to the cubic NLS (on the real line), we have the complex-valued modified Korteweg-de Vries (mKDV) equation

$$\partial_t u + \partial_x^3 u \pm 6|u|^2 \partial_x u = 0, \tag{7.19}$$

with $u(x, 0) = f(x)$. The following global well-posedness result holds, see [191, 192].

Theorem 7.11 *Let* $s \geq 1/4$ *and* $2 \leq p < \infty$. *Then, there exists a function* $C : \mathbb{R}_+ \times \mathbb{R}_+ \to \mathbb{R}_+$, *which is increasing in each argument, such that*

$$\sup\{\|u(t)\|_{\mathscr{M}_s^{2,p}} : t \in [-T, T]\} \leq C(\|f\|_{\mathscr{M}_s^{2,p}}, \tau)$$

for any $\tau > 0$ *and any Schwartz solution* u *to* (7.19). *In particular, this implies that* (7.19) *is globally well posed in* $\mathscr{M}_s^{2,p}$.

Similarly, consider the nonlinear wave equation (NLW)

$$\partial_{tt} u - \Delta u + N(u) = 0, \tag{7.20}$$

and the nonlinear Klein-Gordon equations (NLKG)

$$\partial_{tt}u + (I - \Delta)u + N(u) = 0, \tag{7.21}$$

with initial data $u(x, 0) = f(x)$, $\partial_t u(x, 0) = g(x)$. In each case and as for the NLS, the well posedness is understood as a statement about the corresponding Duhamel formulation.

The following local well-posedness result holds, see [14].

Theorem 7.12 *Let $k \in \mathbb{N} \cup \{0\}$ and N as in (7.18). Then, the NLW (7.20) and NLKG (7.21) are locally well posed in $\mathscr{M}_s^{p,1}$ for all $p \geq 1$ and $s \geq 0$.*

Theorem 7.12 is not the most precise in that we expect a derivative loss of regularity for g. Indeed, this is the case, as it can be seen in the next global result from [155].

Theorem 7.13 *Let k, d, N be as in Theorem 7.8. Define $s = (d + 2)/(d(1 + k))$. Furthermore, let $(f, g) \in \mathscr{M}_s^{2,1} \times \mathscr{M}_{s-1}^{2,1}$ be such that $\|f\|_{\mathscr{M}_s^{2,1}} + \|g\|_{\mathscr{M}_{s-1}^{2,1}} \leq \delta$ for some $0 < \delta < 1$. Then, the NLKG has a unique global solution in $\mathscr{C}(\mathbb{R}; \mathscr{M}_s^{2,1}) \cap \mathscr{Y}_{1+k}$.*

Some of the results presented here can be extended in various ways, in particular by requiring a more general form in the nonlinearity; see, for example, [66] and [28, 29].

7.3.2. There is an incredible amount of activity on the topic of well posedness on a variety of PDEs. Many results, especially in global setting, are extremely difficult to get; they require the full arsenal of "hard analysis" including a deep understanding of time-frequency techniques. Traditionally, one assumes the initial data to be in an appropriate Sobolev space. Depending on the regularity assumed, and generally speaking, below a certain critical one, one can be faced with the problem being ill-posed. For example, a result of Christ-Colliander-Tao [52] states that NLS (7.7) is ill-posed in H^s in the supercritical regime, that is, for $s < d/2 - 2/(p - 1)$. A way to fix this issue is to *randomize* the initial data. Surprisingly enough, it is a randomization adapted to the Wiener decomposition of the frequency space discussed in Sect. 3.2 that comes to the rescue. Indeed, the Wiener randomization intrinsically associated with the modulation spaces is the "right" one from the perspective of the needed time-frequency analysis. Some recent well-posedness results in the realm of probabilistic Cauchy theory that use ideas stemming from the theory of modulation spaces were obtained, for example, in [23, 24] for the NLS and in [179, 190] and [197] for the NLW.

7.3.3. One of the drawbacks of the modulation spaces as function spaces on which to study PDEs is their bias towards one fixed scale on the frequency side. A way to fix this issue is to consider appropriate (weighted) *modulation spaces with scaling symmetry*; see [11]. Some versions of these scale invariant modulation (specifically, Wiener amalgam) spaces have been explored, for example, in connection with the one-dimensional cubic NLS in almost critical spaces in [110]. The scale invariant modulation spaces are related in an appropriate sense with the Morrey spaces; see again [11] and the references therein.

Appendix A
A Proof of Banach's Fixed Point Theorem

Below, we provide a proof of Theorem 7.3.

Let $u_0 \in \mathscr{X}$ be arbitrary and consider the sequence $u_n = T^n(u_0)$ for $n \in \mathbb{N}$. Here, T^n denotes the composition of T with itself n times. First of all, since T is a contraction, a simple induction argument shows that for all $k \in \mathbb{N}$,

$$\|u_{k+1} - u_k\| \leq \lambda^k \|u_1 - u_0\|.$$

Using this fact and the triangle inequality, we have that, for all $m > n$,

$$\|u_m - u_n\| \leq \sum_{j=1}^{m-n} \|u_{j+n} - u_{j+n-1}\|$$

$$\leq \left(\sum_{j=1}^{m-n} \lambda^{j+n-1} \right) \|u_1 - u_0\|$$

$$= \frac{\lambda^n - \lambda^m}{1 - \lambda} \|u_1 - u_0\| \leq \frac{\lambda^n}{1 - \lambda} \|u_1 - u_0\|.$$

This implies that (u_n) is a Cauchy sequence in $\mathscr{X}$. Since $\mathscr{X}$ is complete, it must converge to some limit, say $v \in \mathscr{X}$. Moreover, since T is a contraction, we have

$$\|u_{n+1} - T(v)\| = \|T(u_n) - T(v)\| \leq \lambda \|u_n - v\|.$$

Now, since $\|u_n - v\| \to 0$, we also have $\|u_{n+1} - T(v)\| \to 0$ for $n \to \infty$. By the uniqueness of limit, we must have $T(v) = v$, thus proving that $v \in \mathscr{X}$ is a fixed point for T. To see that v is the unique fixed point of this mapping, assume that $v' \in \mathscr{X}$ is another fixed point of T. By the contraction property,

$$\|v - v'\| = \|T(v) - T(v')\| \leq \lambda \|v - v'\|.$$

Therefore, since $\lambda < 1$, we have $\|v - v'\| = 0$, or $v = v'$.

© Springer Science+Business Media, LLC, part of Springer Nature 2020
Á. Bényi, K. A. Okoudjou, *Modulation Spaces*, Applied and Numerical Harmonic Analysis, https://doi.org/10.1007/978-1-0716-0332-1

Appendix B
The Picard-Lindelöf and Peano Theorems

We present here two results from the theory of ordinary differential equations (ODEs). Let $\Omega \subset \mathbb{R} \times \mathbb{R}^d$ be an open set, $F : \Omega \to \mathbb{R}^d$ a continuous function and $I \subset \mathbb{R}$ an open interval. We say that a function $x : I \to \mathbb{R}^d$ is a solution of the ODE

$$dx/dt = F(t, x) \tag{B.1}$$

if for all $t \in I$ we have $(t, x(t)) \in \Omega$, $x \in \mathscr{C}^1(I)$ and $dx/dt = F(t, x(t))$. Moreover, if $t_0 \in I$, $x(t_0) = x_0$, and x is a solution of (B.1), we say that x is a solution of the associated initial value (or Cauchy) problem with initial data (t_0, x_0).

It is straightforward to prove that given $(t_0, x_0) \in \Omega$, $x : I \to \mathbb{R}^d$ is a solution of the Cauchy problem (B.1) with initial condition (t_0, x_0) if and only if for all $t \in I$,

$$x(t) = x_0 + \int_{t_0}^{t} F(s, x(s)) \, ds. \tag{B.2}$$

We say that $F : \Omega \to \mathbb{R}^d$ is *locally Lipschitz in x* if for any compact set $\mathbb{K} \subset \Omega$, there exists a constant $\gamma = \gamma(\mathbb{K}, F) > 0$ such that for all $(t, x), (t, y) \in \mathbb{K}$, we have

$$|F(t, x) - F(t, y)| \leq \gamma \, |x - y|.$$

The next theorem provides a sufficient condition for the existence and uniqueness of (local) solutions of an ODE.

Theorem B.1 (Picard-Lindelöf) *If $F : \Omega \to \mathbb{R}^d$ is both continuous on Ω and locally Lipschitz in x, then for each $(t_0, x_0) \in \Omega$ there exists a unique solution $x : I \to \mathbb{R}^d$ of (B.1) with $x(t_0) = x_0$, where I is some open interval containing t_0. Moreover, for every $\epsilon > 0$, there exists $\mu > 0$ such that if $|x_0 - \tilde{x}_0| < \mu$ and $\tilde{x}$ is a solution of (B.1) with $\tilde{x}(t_0) = \tilde{x}_0$, then $|x(t) - \tilde{x}(t)| < \epsilon$ for all $t \in I$.*

© Springer Science+Business Media, LLC, part of Springer Nature 2020
Á. Bényi, K. A. Okoudjou, *Modulation Spaces*, Applied and Numerical
Harmonic Analysis, https://doi.org/10.1007/978-1-0716-0332-1

Proof Given some interval $I \subset \mathbb{R}$, we write $\mathscr{C}(I)$ for the set of continuous and bounded functions $x : I \to \mathbb{R}^d$. We endow this space with the complete norm

$$\|x\| = \sup_{t \in I} |x(t)|,$$

thus making $\mathscr{C}(I)$ a Banach space. Since Ω is open and $(t_0, x_0) \in \Omega$, there exist some (fixed) $a, b \in \mathbb{R}$ and $r > 0$ such that $t_0 \in (a, b)$ and the compact set $\mathbb{K} = [a, b] \times \overline{\mathbb{B}(x_0, r)} \subset \Omega$. Define

$$c = c(\mathbb{K}, F) = \sup\{|F(t, x)| : (t, x) \in \mathbb{K}\}.$$

Since F is continuous on $\mathbb{K}$, we have $c \in [0, \infty)$.

Recall that finding a solution of (B.1) with $x(t_0) = x_0$ is equivalent to finding an $x \in \mathscr{C}(I)$, $I \ni t_0$, such that (B.2) is satisfied for all $t \in I$. Consider $\delta > 0$ sufficiently small with $I := (t_0 - \delta, t_0 + \delta) \subset (a, b)$ and such that $c\delta < r$ and $\gamma\delta < 1$. Denote by $\mathscr{X}$ the closed ball in $\mathscr{C}(I)$ centered at x_0 and of radius r, that is,

$$\mathscr{X} := \{y \in \mathscr{C}(I) : \|y - x_0\| \leq r\}.$$

Note that $\mathscr{X}$ is nonempty, and because $\mathscr{C}(I)$ is a Banach space so is $\mathscr{X}$ with the induced norm. Let now $T : \mathscr{X} \to \mathscr{X}$ be defined by

$$T(x)(t) := x_0 + \int_{t_0}^t F(s, x(s))\, ds, \quad x \in \mathscr{X}, t \in I.$$

First of all, T is well defined on $\mathscr{X}$ because $x \in \mathscr{X}$ implies that $|x(s) - x_0| \leq r$ for all $s \in I$, that is $x(s) \in \overline{\mathbb{B}(x_0, r)}$ for all $s \in I$. Moreover, if $x \in \mathscr{X}$ we have $T(x) \in \mathscr{C}(I)$ and for all $t \in I$ we have

$$|T(x)(t) - x_0| \leq \int_{t_0}^t |F(s, x(s))|\, ds \leq c|t - t_0| \leq c\delta < r.$$

Thus, T is indeed a transformation from $\mathscr{X}$ to $\mathscr{X}$.

Next, we show that T is a contraction on $\mathscr{X}$. Let then $x, y \in \mathscr{X}$. Using the fact that F is locally Lipschitz in x, we have, for all $t \in I$,

$$|T(x)(t) - T(y)(t)| \leq \int_{t_0}^t |F(s, x(s)) - F(s, y(s))|\, ds$$

$$\leq \gamma \int_{t_0}^t |x(s) - y(s)|\, ds \leq \gamma\delta\|x - y\|,$$

or

$$\|T(x) - T(y)\| \leq \lambda\|x - y\|,$$

where $\lambda = \gamma\delta < 1$. Since T is a contraction on the Banach space $\mathscr{X}$, by Banach's fixed point theorem, T has a unique fixed point $x \in \mathscr{X}$, which is the solution of our ODE with initial data (t_0, x_0).

For the second part of the theorem, we show that we have a Lipschitz dependence of the solutions on the initial data. More precisely, we claim that there exists some constant $C > 0$ such that

$$\|x - \widetilde{x}\| \leq C|x_0 - \widetilde{x}_0|. \tag{B.3}$$

Indeed, using again (B.2) and the locally Lipschitz condition in x of F, we see that for all $t \in I$,

$$|x(t) - \widetilde{x}(t)| \leq |x_0 - \widetilde{x}_0| + \int_{t_0}^{t} |F(s, x(s)) - F(s, \widetilde{x}(s))|\, ds$$

$$\leq |x_0 - \widetilde{x}_0| + \gamma \int_{t_0}^{t} |x(s) - \widetilde{x}(s)|\, ds.$$

From here, using Gronwall's lemma, see Appendix C, we get

$$|x(t) - \widetilde{x}(t)| \leq |x_0 - \widetilde{x}_0|e^{\gamma(t-t_0)} \leq e^{\gamma\delta}|x_0 - \widetilde{x}_0|.$$

Thus, letting $C = e^{\gamma\delta}$, we obtain (B.3); this clearly proves the second part of the theorem, by simply letting $\mu = \epsilon/C$. $\square$

Remark B.2 The Picard-Lindelöf theorem is in effect proving the *local well posedness of* (B.1) under the additional assumption on F being locally Lipschitz in x. It also turns out that each solution of the Cauchy problem given by this theorem can be extended uniquely to a maximal interval of existence.

Motivated by the previous remark, one can naturally ask whether the additional assumption on F can be removed. The answer is yes, as long as one is concerned only with the existence of solutions (but not uniqueness!). The corresponding result is due to Peano, and its proof is based on the theorem of Arzelá-Ascoli which characterizes the relatively compact sets of continuous functions on a Hausdorff space (in the topology induced by the uniform norm) in terms of equicontinuity and pointwise boundedness; see, for example, [2, Theorem 1.49] for further details.

Theorem B.3 (Peano) *If F is continuous on Ω, then for each $(t_0, x_0) \in \Omega$ there exists at least one solution $x : I \rightarrow \mathbb{R}^d$ of (B.1), where I is an open interval such that $t_0 \in I$ and $x(t_0) = x_0$.*

Appendix C
Gronwall's Lemma

The dependence of solutions on the initial data, say in the Picard-Lindelöf theorem, is essentially dependent on a classical lemma due to Gronwall, which we now state and prove; see [2, Proposition 1.39].

Lemma C.1 (Gronwall) *Let $c \in \mathbb{R}$ be a constant and $u, v : [a, b] \to \mathbb{R}^d$ two continuous functions with $v \geq 0$ and $u(t) \leq c + \int_a^t u(s)v(s) \, ds$ for all $t \in [a, b]$. Then, for all $t \in [a, b]$, we have*

$$u(t) \leq c e^{\int_a^t v(s) \, ds}.$$

Proof Let $F(t) = \int_a^t u(s)v(s) \, ds$ and $G(t) = \int_a^t v(s) \, ds$. The inequality in the hypothesis implies that

$$F'(t) - v(t)F(t) \leq cv(t).$$

Equivalently,

$$\left(e^{-G(t)} F(t)\right)' = e^{-G(t)}(F'(t) - v(t)F(t)) \leq cv(t)e^{-G(t)}.$$

Integrating both sides of this inequality between a and t and recalling that $F(a) = 0$ and $G'(t) = v(t)$, we obtain

$$e^{-G(t)} F(t) \leq c(1 - e^{-G(t)}) \Leftrightarrow F(t) \leq ce^{G(t)} - c.$$

Finally, since $u(t) \leq c + F(t)$, we get $u(t) \leq ce^{G(t)}$ for all $t \in [a, b]$. This is exactly what we wanted to prove. □

© Springer Science+Business Media, LLC, part of Springer Nature 2020
Á. Bényi, K. A. Okoudjou, *Modulation Spaces*, Applied and Numerical Harmonic Analysis, https://doi.org/10.1007/978-1-0716-0332-1

Appendix D
Local Well Posedness of NLS on Sobolev Spaces

In what follows, we discuss the "folklore" version of local well posedness of the nonlinear Schrödinger equation (7.7) with initial data f on $H^s(\mathbb{R}^d)$, $s > d/2$, which in the Duhamel formulation (7.8) is equivalent to proving that the operator T defined in (7.12) by

$$T(u)(t) = S(t)(f) - i \int_0^t S(t - t')N(u(t')) \, dt',$$

with $t \in [-\tau, \tau]$ has a (unique) fixed point $u(t) \in H^s := H^s(\mathbb{R}^d)$. Here, recall that $S(t) = e^{it\Delta}$ denotes the linear Schrödinger operator and $N(u) = \pm|u|^{k-1}u$; for the simplicity of the exposition, we assume $k \in \mathbb{N}$, $k \geq 2$. We also write, for

$$F(t) : \mathbb{R}^d \to \mathbb{C}, \ F(t)(x) = F(t, x),$$

$F \in \mathscr{C}_t H^s := \mathscr{C}(\mathbb{R}; H^s)$ if $F(t) \in H^s(\mathbb{R}^d)$ and $\|F(t + h) - F(t)\|_{H^s} \to 0$ as $h \to 0$ for all $t \in \mathbb{R}$. Note that if $F \in \mathscr{C}_t H^s$, then $\|F(t)\|_{H^s} \in \mathscr{C}(\mathbb{R}; \mathbb{R}_+)$. We define the norm on $\mathscr{C}_t H^s$ by

$$\|F\|_{\mathscr{C}_t H^s} := \left\| \|F(t)\|_{H^s(\mathbb{R}_x^d)} \right\|_{L^\infty(\mathbb{R}_t)}.$$

For some fixed $\tau > 0$, we also write $F \in \mathscr{C}_\tau H^s$ if $F \in \mathscr{C}([-\tau, \tau]; H^s)$.

Before considering the question of well posedness, let us first record a few useful observations.

(i) *For all $s \in \mathbb{R}$, $\{S(t)\}_{t \in \mathbb{R}}$ is a unitary group on H^s.* That is, $S(t + t') = S(t)S(t')$ for $t, t' \in \mathbb{R}$—which follows immediately from the definition of $S(t)$ as a Fourier multiplier operator—and $\|S(t)(f)\|_{H^s} = \|f\|_{H^s}$—which is a consequence of Plancherel's identity.

(ii) *If the initial data $f \in H^s$, then $S(t)(f) \in \mathscr{C}_t H^s$.*

© Springer Science+Business Media, LLC, part of Springer Nature 2020
Á. Bényi, K. A. Okoudjou, *Modulation Spaces*, Applied and Numerical Harmonic Analysis, https://doi.org/10.1007/978-1-0716-0332-1

Let us fix $f \in H^s$, $f \not\equiv 0$, and $t \in \mathbb{R}$. Then, for all $h \in \mathbb{R}$, the unitary group property of $S(t)$ gives

$$
\begin{aligned}
\|S(t+h)(f) - S(t)(f)\|_{H^s}^2 &= \|S(h)(f) - f\|_{H^s}^2 \\
&= \int_{\mathbb{R}^d} \langle \xi \rangle^{2s} |e^{-4\pi^2 i h |\xi|^2} - 1|^2 |\widehat{f}(\xi)|^2 \, d\xi.
\end{aligned}
\tag{D.1}
$$

Let now $\epsilon > 0$ be arbitrary. Since $\langle \xi \rangle^s \widehat{f}(\xi) \in L^2(\mathbb{R}^d)$, there exists $N = N(\epsilon, f) > 0$ such that

$$
\int_{|\xi| > N} \langle \xi \rangle^{2s} |\widehat{f}(\xi)|^2 \, d\xi < \frac{\epsilon^2}{16};
$$

in particular, this implies that

$$
\int_{|\xi| > N} \langle \xi \rangle^{2s} |e^{-4\pi^2 i h |\xi|^2} - 1|^2 |\widehat{f}(\xi)|^2 \, d\xi < \frac{\epsilon^2}{4}.
\tag{D.2}
$$

Now, by the mean value theorem we have

$$
|e^{-4\pi^2 i h |\xi|^2} - 1|^2 \le 4\pi^2 |\xi|^2 |h|^2,
$$

which further yields

$$
\int_{|\xi| \le N} \langle \xi \rangle^{2s} |e^{-4\pi^2 i h |\xi|^2} - 1|^2 |\widehat{f}(\xi)|^2 \, d\xi \le 4\pi^2 N^2 |h|^2 \|f\|_{H^s}^2.
\tag{D.3}
$$

Letting $0 < |h| < \epsilon/(4\pi N \|f\|_{H^s})$, (D.1)–(D.3) yield

$$
\|S(t+h)(f) - S(t)(f)\|_{H^s}^2 < \frac{\epsilon^2}{4} + \frac{\epsilon^2}{4} < \epsilon^2;
$$

this proves that, in fact, $S(t)(f)$ is a uniformly continuous function in t with values in H^s.

(iii) *For any $F \in \mathscr{C}_t H^s$, we have $\int_0^t S(t - t') F(t') \, dt' \in \mathscr{C}_t H$.*

The argument proving this fact is similar to the one in (ii) above. Let us write

$$
G(t) = \int_0^t S(t - t') F(t') \, dt'
$$

and $I = S(0)$ for the identity operator. Fix $t > 0$ and $F \not\equiv 0$. Using again the group property of $S(t)$ and the linearity of the integral we see that

$$G(t+h)-G(t) = \int_0^t (S(h)-I)S(t-t')F(t')\,dt' + \int_t^{t+h} S(t+h-t')F(t')\,dt'.$$

Then, by the unitary property of $S(t)$ on H^s, we obtain

$$\|G(t+h)-G(t)\|_{H^s} \le \|(S(h)-I)(G)\|_{\mathscr{C}(\mathbb{R};H^s)} + |h|\|F\|_{\mathscr{C}(\mathbb{R};H^s)}.$$

Thus, choosing $0 < |h| < \epsilon/(2\|F\|_{\mathscr{C}_t H^s})$ and using (ii), we obtain that

$$\|G(t+h)-G(t)\|_{H^s} < \epsilon,$$

which proves our claim.

We have now the necessary tools to prove that our operator T *has a unique fixed point in the closed ball* $\overline{\mathbb{B}(0,R)} \subset \mathscr{C}_\tau H^s$, $s > d/2$, where

$$R = 2\|f\|_{H^s} \text{ and } \tau \le \frac{1}{2R^{k-1}} \sim \|f\|_{H^s}^{1-k}.$$

We will show that $T : \overline{\mathbb{B}(0,R)} \to \overline{\mathbb{B}(0,R)}$ and that T is a contraction on $\overline{\mathbb{B}(0,R)}$.

Fix $f \in H^s$. First, by the unitary property of $S(t)$, Minkowski's integral inequality and the fact that H^s is an algebra for $s > d/2$, we have the following:

$$\|T(u)\|_{\mathscr{C}_\tau H^s} \le \|S(t)(f)\|_{\mathscr{C}_\tau H^s} + \int_0^\tau \left\||u|^{k-1}u\right\|_{\mathscr{C}_\tau H^s} dt'$$

$$\le \|f\|_{H^s} + \tau\|u\|_{\mathscr{C}_\tau H^s}^k \le 2\|f\|_{H^s} = R;$$

here, we used the fact that $u \in \overline{\mathbb{B}(0,R)}$ gives

$$\tau\|u\|_{\mathscr{C}_\tau H^s}^k \le \tau R^k \le R/2$$

for $\tau \le 1/(2R^{k-1})$.

Second, for $u, v \in \overline{\mathbb{B}(0,R)} \subset \mathscr{C}_\tau H^s$, using again Minkowski's inequality and the mean value theorem (or, appropriately telescoping the difference $|u|^{k-1}u - |v|^{k-1}v$, say, if k is odd), we have

$$\|T(u)-T(v)\|_{\mathscr{C}_\tau H^s} \le \int_0^\tau \left\||u|^{k-1}u - |v|^{k-1}v\right\|_{\mathscr{C}_\tau H^s} dt'$$

$$\le c\tau(\|u\|_{\mathscr{C}_\tau H^s}^{k-1} + \|v\|_{\mathscr{C}_\tau H^s}^{k-1})\|u-v\|_{\mathscr{C}_\tau H^s},$$

for some appropriate constant $c > 0$. Thus, choosing now τ sufficiently small such that

$$c\tau(\|u\|_{\mathscr{C}_\tau H^s}^{k-1} + \|v\|_{\mathscr{C}_\tau H^s}^{k-1}) < 1,$$

we obtain that T is a contraction on $\overline{\mathbb{B}(0, R)}$. Therefore, by Banach's fixed point theorem, we conclude that there exists a unique fixed point $u \in \overline{\mathbb{B}(0, R)} \subset \mathscr{C}_\tau H^s$ such that $T(u) = u$, which is our (local-in-time) solution to the NLS.

It is worth noting that, using a so-called *bootstrap argument*, it can be shown that the local-in-time solution u is in fact unique in the entire $\mathscr{C}_\tau H^s$ by possibly shrinking the existence time τ by a constant factor which does not change the order of the existence and uniqueness time; that is, given $f \in H^s$, if $u, v \in \mathscr{C}_\tau H^s$ are two solutions of the NLS with $u(0) = v(0) = f$, then $u = v$ and $\tau \sim \|f\|_{H^s}^{1-k}$.

Roughly speaking, the bootstrap argument allows us to control the size of a solution $u \in \mathscr{C}_\tau H^s$ to the NLS; more precisely, given $f \in H^s$ and $u(0) = f$, we can in fact show that $u \in \overline{\mathbb{B}(0, R)} \subset \mathscr{C}_{\tau_0} H^s$ for some $\tau_0 = c_0 R^{1-k}$.

Finally, let us also mention that an argument similar to the one proving the contraction property of T can be used to prove that there is a well-defined *solution map* which is Lipschitz continuous on H^s. For further details, the interested reader can consult for example [189, pp. 10–11].

References

1. Balan, R., Christensen, J., Krishtal, I., Okoudjou, K.A., Romero, J.-L.: Multi-window Gabor frames in amalgam spaces. Math. Res. Lett. **21**(1), 1–15 (2014)
2. Barreira, L., Valls, C.: Ordinary differential equations. In: Qualitative Theory. Graduate Studies in Mathematics, vol. 137. American Mathematical Society, Providence (2012)
3. Beckner, W.: Inequalities in Fourier analysis. Ann. Math. **102**, 159–182 (1975)
4. Benedetto, J.J.: Harmonic analysis and applications. In: Studies in Advanced Mathematics. CRC, Boca Raton (1997)
5. Benedetto, J.J., Czaja, W.: Integration and modern analysis. In: Birkhäuser Advanced Texts: Basel Textbooks. Birkhäuser, Boston (2009)
6. Bennett, C., Sharpley, R.: Interpolation of operators. In: Pure and Applied Mathematics, vol. 129. Academic, Boston (1988)
7. Bényi, Á.: Bilinear pseudodifferential operators on Lipschitz and Besov spaces. J. Math. Anal. Appl. **284**, 97–103 (2003)
8. Bényi, Á., Oh, T.: Modulation spaces, Wiener amalgam spaces, and Brownian motions. Adv. Math. **228**(5), 2943–2981 (2011)
9. Bényi, Á., Oh, T.: The Sobolev inequality on the torus revisited. Publ. Math. Debr. **83**(3), 359–374 (2013)
10. Bényi, Á., Oh, T.: On a class of bilinear pseudodifferential operators. J. Funct. Spaces Appl. **2013**, 560976 (2013)
11. Bényi, Á., Oh, T.: Modulation spaces with scaling symmetry. Appl. Comput. Harmon. Anal. **48**(1), 496–507 (2020)
12. Bényi, Á., Okoudjou, K.A.: Modulation space estimates for multilinear pseudodifferential operators. Stud. Math. **172**(2), 169–180 (2006)
13. Bényi, Á., Okoudjou, K.A.: Time-frequency estimates for pseudodifferential operators. In: Harmonic Analysis, Partial Differential Equations, and Related Topics. Contemporary Mathematics, vol. 428, pp. 13–22. American Mathematical Society, Providence (2007)
14. Bényi, Á., Okoudjou, K.A.: Local well-posedness of nonlinear dispersive equations on modulation spaces. Bull. Lond. Math. Soc. **41**, 549–558 (2009)
15. Bényi, Á., Torres, R.H.: Symbolic calculus and the transposes of bilinear pseudodifferential operators. Comm. Partial Differ. Equ. **28**, 1161–1181 (2003)
16. Bényi, Á., Torres, R.H.: Almost orthogonality and a class of bounded bilinear pseudodifferential operators. Math. Res. Lett. **11**, 1–11 (2004)
17. Bényi, Á., Grafakos, L., Gröchenig, K., Okoudjou, K.A.: A class of Fourier multipliers for modulation spaces. Appl. Comput. Harmon. Anal. **19**(1), 131–139 (2005)

18. Bényi, Á., Gröchenig, K., Heil, C., Okoudjou, K.A.: Modulation spaces and a class of bounded multilinear pseudodifferential operators. J. Oper. Theory **54**, 301–313 (2005)
19. Bényi, Á., Nahmod, A.R., Torres, R.H.: Sobolev space estimates and symbolic calculus for bilinear pseudodifferential operators. J. Geom. Anal. **16**(3), 431–453 (2006)
20. Bényi, Á., Gröchenig, K., Okoudjou, K.A., Roger, L.G: Unimodular Fourier multipliers for modulation spaces. J. Funct. Anal. **246**(2), 366–384 (2007)
21. Bényi, Á., Maldonado, D., Naibo, V., Torres, R.H.: On the Hörmander classes of bilinear pseudodifferential operators. Integr. Equ. Oper. Theory **67**(3), 341–364 (2010)
22. Bényi, Á., Bernicot, F., Maldonado, D., Naibo, V., Torres, R.H.: On the Hörmander classes of bilinear pseudodifferential operators, II. Indiana Univ. Math. J. **62**(6), 1733–1764 (2013)
23. Bényi, Á., Oh, T., Pocovnicu, O.: On the probabilistic Cauchy theory of the cubic nonlinear Schrödinger equation on $\mathbb{R}^d$, $d \geq 3$. Trans. Am. Math. Soc. Ser. B **2**, 1–50 (2015)
24. Bényi, Á., Oh, T., Pocovnicu, O.: Wiener randomization on unbounded domains and an application to almost sure well-posedness of NLS. In: Excursions in Harmonic Analysis. Applied and Numerical Harmonic Analysis, vol. 4, pp. 3–25. Birkhäuser/Springer, Berlin (2015)
25. Berezin, F.A.: Wick and anti-Wick symbols of operators. Math. Sb. **86**(128), 578–610 (1971)
26. Bertrandias, J.-P., Datry, C., Dupuis, C.: Unions et intersections d'espaces L^p invariantes par translation ou convolution. Ann. Inst. Fourier (Grenoble) **28**(2), 53–84 (1978)
27. Beurling, A., Helson, H.: Fourier-Stieltjes transform with bounded powers. Math. Scand. **1**, 12–126 (1953)
28. Bhimani, D.G., Ratnakumar, P.K.: Functions operating on modulation spaces and nonlinear dispersive equations. J. Funct. Anal. **270**(2), 621–648 (2016)
29. Bhimani, D.G., Ratnakumar, P.K.: Erratum to "Functions operating on modulation spaces and nonlinear dispersive equations". J. Funct. Anal. **270**(6), 2375 (2016)
30. Boggiatto, P., Cordero, E.: Anti-Wick quantization with symbols in L^p spaces. Proc. Am. Math. Soc. **130**(9), 2679–2685 (2002)
31. Boggiatto, P., Toft, J.: Schatten classes for Toeplitz operators with Hilbert space windows on modulation spaces. Adv. Math. **217**(1), 305–333 (2008)
32. Boggiatto, P., Cordero, E., Gröchenig, K.: Generalized anti-Wick operators with symbols in distributional Sobolev spaces. Integr. Equ. Oper. Theory **48**(4), 427–442 (2004)
33. Boggiatto, P., Oliaro, A., Wong, M.W.: L^p-boundedness and compactness of localization operators. J. Math. Anal. Appl. **322**(1), 193–206 (2006)
34. Bony, J.-M.: Calcul symbolique et propagation des singularités pour les équations aux dérivées partielles non linéaires. Ann. Sci. École Norm. Sup. **14**(4), 209–246 (1981)
35. Boulkhemair, A.: Remarks on a Wiener type pseudodifferential algebra and Fourier integral operators. Math. Res. Lett. **4**, 53–67 (1997)
36. Calderón, A.P.: Cauchy integrals of Lipschitz curves and related operators. Proc. Natl. Acad. Sci. USA **74**, 1324–1397 (1977)
37. Calderón, A.P.: Commutators of singular integral operators. Proc. Natl. Acad. Sci. USA **53**, 1092–1099 (1977)
38. Calderón, A.P., Vaillancourt, R.: On the boundedness of pseudodifferential operators. J. Math. Soc. Jpn. **23**, 374–378 (1971)
39. Calderón, A.P., Vaillancourt, R.: A class of bounded pseudodifferential operators. Proc. Natl. Acad. Sci. USA **69**, 1185–1187 (1972)
40. Calderón, A.P., Zygmund, A.: On the existence of certain singular integrals. Acta Math. **88**, 85–139 (1952)
41. Carleson, L.: On convergence and growth of partial sums of Fourier series. Acta Math. **116**, 135–157 (1966)
42. Cazenave, T., Weissler, F.: Some remarks on the nonlinear Schrödinger equation in the critical case. In: Nonlinear Semigroups, Partial Differential Equations and Attractors (Washington, DC, 1987). Lecture Notes in Mathematical, vol. 1394, pp. 18–29. Springer, Berlin (1989)
43. Chen, M., Guo, B.: Local Well and Ill Posedness for the Modified KdV Equations in Subcritical Modulation Spaces (2018). arXiv:1811.05182v1

44. Chen, J.C., Zhong, Y.: Modulation space estimates for the fractional integral operators. Sci. China Math. **54**(7), 1478–1489 (2011)
45. Chen, J.C., Fan, D.S., Sun, L.J.: Asymptotic estimates for unimodular multiplies on modulation spaces. Discrete Contin. Dynam. Syst. **32**(2), 467–485 (2012)
46. Chen, J.C., Fan, D.S., Sun, L.J., Zhang, C.J.: Estimates for unimodular multiplies on modulation Hardy spaces. J. Funct. Spaces Appl. **2013**, 16 (2013). Art. ID 982753
47. Chen, J.C., Guo, W.C., Zhao, G.: Remarks on the unimodular Fourier multipliers on α-modulation spaces. J. Funct. Spaces **2014**, 8 (2014). Art. ID 106267
48. Chen, J.C., Fan, D., Guo, W.C., Zhao, G.: Sharp estimates of unimodular multipliers on frequency decomposition spaces. Nonlinear Anal. **142**, 26–47 (2016)
49. Chen, J.C., Fan D.S., Huang, Q.: Critical exponent for evolution equations in modulation spaces. J. Math. Anal. Appl. **443**(1), 230–242 (2016)
50. Chen, J.C., Huang, Q., Zhu, X.R.: Non-integer power estimate in modulation spaces and its applications. Sci. China Math. **60**(8), 1443–1460 (2017)
51. Christ, M., Weinstein, M.: Dispersion of small amplitude solutions of the generalized Korteweg-de Vries equation. J. Funct. Anal. **100**, 87–109 (1991)
52. Christ, M., Colliander, J., Tao, T.: Asymptotics, frequency modulation, and low-regularity ill-posedness of canonical defocusing equations. Am. J. Math. **125**(6), 1235–1293 (2003)
53. Christensen, O.: An introduction to frames and Riesz bases. In: Applied and Numerical Harmonic Analysis. Birkäuser, Boston (2003)
54. Coifman, R.R., Meyer, Y.: On commutators of singular integrals and bilinear singular integrals. Trans. Am. Math. Soc. **212**, 315–331 (1975)
55. Coifman, R.R., Meyer, Y.: Commutateurs d'intégrales singulières et opérateurs multilinéaires. Ann. Inst. Fourier **28**, 177–202 (1978)
56. Coifman, R.R., Meyer, Y.: Au delà des opérateurs pseudo-différentiels. In: Astérisque, vol. 57. Société Mathématique de France, Paris (1978)
57. Coifman, R.R., Meyer, Y.: Nonlinear harmonic analysis, operator theory and P.D.E. In: Beijing Lectures in Harmonic Analysis (Beijing, 1984). Annals of Mathematics Studies, vol. 112, pp. 3–45. Princeton University Press, Princeton (1986)
58. Coifman, R.R., Meyer, Y.: Wavelets: Calderón-Zygmund and Multilinear Operators. Cambridge University Press, Cambridge (1997)
59. Concetti, F., Toft, J.: Schatten-von Neuman properties for Fourier integral operators with non-smooth symbols. I. Ark. Mat. **47**, 295–312 (2009)
60. Cordero, E., Gröchenig, K.: Time-frequency analysis of localization operators. J. Funct. Anal. **205**(1), 107–131 (2003)
61. Cordero, E., Gröchenig, K.: Necessary conditions for Schatten class localization operators. Proc. Am. Math. Soc. **133**(12), 3573–3579 (2005)
62. Cordero, E., Gröchenig, K.: Symbolic calculus and Fredholm property for localization operators. J. Fourier Anal. Appl. **12**(4), 371–392 (2006)
63. Cordero, E., Gröchenig, K.: On the product of localization operators. In: Modern Trends in Pseudo-Differential Operators. Operator Theory: Advances and Applications, vol. 172, pp. 279–295. Birkhäuser, Basel (2007)
64. Cordero, E., Nicola, F.: Some new Strichartz estimates for the Schrödinger equation. J. Differ. Equ. **245**(7), 1945–1974 (2008)
65. Cordero, E., Nicola, F.: Strichartz estimates in Wiener amalgam spaces for the Schrödinger equation. Math. Nachr. **281**(1), 25–41 (2008)
66. Cordero, E., Nicola, F.: Remarks on Fourier multipliers and applications to the wave equation. J. Math. Anal. Appl. **353**(2), 583–591 (2009)
67. Cordero, E., Nicola, F.: Sharp continuity results for the short-time Fourier transform and for localization operators. Monatsh. Math. **162**(3), 251–276 (2011)
68. Cordero, E., Nicola, F.: On the Schrödinger equation with potential in modulation spaces. J. Pseudo-Differ. Oper. Appl. **5**(3), 319–341 (2014)
69. Cordero, E., Nicola, F.: Sharp integral bounds for Wigner distributions. Int. Math. Res. Not. **6**, 177–1807 (2018)

70. Cordero, E., Okoudjou, K.A.: Multilinear localization operators. J. Math. Anal. Appl. **325**(2), 1103–1116 (2007)
71. Cordero, E., Rodino, L.: Wick calculus: a time-frequency approach. Osaka J. Math. **42**(1), 43–63 (2005)
72. Cordero, E., Rodino, L.: Short-time Fourier transform analysis of localization operators. In: Frames and Operator Theory in Analysis and Signal Processing. Contemporary Mathematics, vol. 451, pp. 47–68. American Mathematical Society, Providence (2008)
73. Cordero, E., Pilipovic, S., Rodino, L., Teofanov, N.: Localization operators and exponential weights for modulation spaces. Mediterr. J. Math. **2**(4), 381–394 (2005)
74. Cordero, E., Gröchenig, K., Rodino, L.: Localization operators and time-frequency analysis. In: Harmonic, Wavelet and p-adic Analysis, pp. 83–110. World Science Publication, Hackensack (2007)
75. Cordero, E., Feichtinger, H., Luef, F.: Banach Gelfand triples for Gabor analysis. In: Pseudo-Differential Operators. Lecture Notes in Mathematics, vol. 1949, pp. 1–33. Springer, Berlin (2008)
76. Cordero, E., Nicola, F., Rodino, L.: Boundedness of Fourier integral operators on $\mathscr{F}L^p$ spaces. Trans. Am. Math. Soc. **361**(11), 6049–6071 (2009)
77. Cordero, E., Nicola, F., Rodino, L.: Time-frequency analysis of Fourier integral operators. Commun. Pure Appl. Anal. **9**, 1–21 (2010)
78. Cordero, E., Gröchenig, K., Nicola, F.: Approximation of Fourier integral operators by Gabor multipliers. J. Fourier Anal. Appl. **18**(4), 661–684 (2012)
79. Cordero, E., Nicola, F., Rodino, L.: Time-frequency analysis of Schrödinger propagators. In: Evolution Equations of Hyperbolic and Schrödinger Type. Progress in Mathematics, vol. 301, pp. 63–85. Birkhäuser/Springer, Basel (2012)
80. Cordero, E., Gröchenig, K., Nicola, F., Rodino, L.: The Wiener property for a class of Fourier integral operators. J. Math. Pures Appl. **99**(2), 219–233 (2013)
81. Cordero, E., Tabacco, A., Wahlberg, P.: Schrödinger type propagators, pseudodifferential operators and modulation spaces. J. Lond. Math. Soc. **88**(2), 375–395 (2013)
82. Cordero, E., Nicola, F., Rodino, L.: Schrödinger equations in modulation spaces. In: Studies in Phase Space Analysis with Applications to PDEs. Program Nonlinear Differential Equations Application, vol. 84, pp. 81–99. Birkhäuser/Springer, New York (2013)
83. Cordero, E., Nicola, F., Rodino, L.: Schrödinger equations with rough Hamiltonians. Discrete Contin. Dyn. Syst. **35**(10), 4805–4821 (2015)
84. Cowling, M.G., Price, J.F.: Bandwidth versus time concentration: the Heisenberg-Pauli-Weyl inequality. SIAM J. Math. Anal. **15**(1), 151–165 (1984)
85. Dahlberg, B.: Estimates of harmonic measure. Arch. Ration. Mech. Anal. **65**, 275–288 (1977)
86. Dahlberg, B.: On the Poisson integral for Lipschitz and C^1 domanins. Studia Math. **66**, 13–24 (1979)
87. Daubechies, I.: Time-frequency localization operators: a geometric phase approach. IEEE Trans. Inform. Theory **34**(4), 605–612 (1988)
88. Daubechies, I.: Ten Lectures on Wavelets. Society for Industrial and Applied Mathematics, Philadelphia (1992)
89. Daubechies, I., Grossmann, A., Meyer, Y.: Painless nonorthogonal expansions. J. Math. Phys. **27**(5), 1271–1283 (1986)
90. Deng, Q., Ding, Y., Sun, L.: Estimate for generalized unimodular multipliers on modulation spaces. Nonlinear Anal. **85**, 78–92 (2013)
91. Dörfler, M., Feichtinger, H.G., Gröchenig, K.: Compactness criteria in function spaces. Colloq. Math. **94**(1), 37–50 (2002)
92. Fefferman, C.: Pointwise convergence of Fourier series. Ann. Math. **98**, 551–571 (1973)
93. Feichtinger, H.G.: On a new Segal algebra. Monatsh. Math. **92**(4), 269–289 (1981)
94. Feichtinger, H.G.: Modulation spaces on locally compact abelian groups. In: Radha, R., Krishna, M., Thangavelu, S. (eds.) Proceeding of International Conference on Wavelets and Applications (Chennai, 2002), pp. 1–56. New Delhi Allied, New Delhi (2003). Technical report, University of Vienna (1983)

95. Feichtinger, H.G.: Generalized amalgams, with applications to Fourier transform. Can. J. Math. **42**(3), 395–409 (1990)
96. Feichtinger, H.G.: Modulation spaces: looking back and ahead. Sampl. Theory Signal Image Process. **5**(2), 109–140 (2006)
97. Feichtinger, H.G.: Choosing function spaces in harmonic analysis. In: Excursions in Harmonic Analysis. Applied and Numerical Harmonic Analysis, vol. 4, pp. 65–101. Birkhäuser/Springer, Cham (2015)
98. Feichtinger, H.G.: Thoughts on numerical and conceptual harmonic analysis. In: New Trends in Applied Harmonic Analysis. Applied and Numerical Harmonic Analysis, pp. 301–329. Birkhäuser/Springer, Berlin (2016)
99. Feichtinger, H.G., Gröchenig, K.: A unified approach to atomic decompositions via integrable group representations. In: Function Spaces and Applications (Lund, 1986). Lecture Notes in Mathematics, vol. 1302, pp. 52–73. Springer, Berlin (1988)
100. Feichtinger, H.G., Gröchenig, K.: Banach spaces related to integrable group representations and their atomic decompositions, I. J. Funct. Anal. **86**(2), 307–340 (1989)
101. Feichtinger, H.G., Gröchenig, K.: Gabor frames and time-frequency analysis of distributions. J. Funct. Anal. **146**(2), 464–495 (1997)
102. Feichtinger, H.G., Narimani, G.: Fourier multipliers of classical modulation spaces. Appl. Comput. Harmon. Anal. **21**(3), 349–359 (2006)
103. Feichtinger, H.G., Strohmer, T.: Gabor analysis and algorithms. In: Applied and Numerical Harmonic Analysis. Birkhäuser, Boston (1998)
104. Feichtinger, H.G., Strohmer, T.: Advances in Gabor analysis. In: Applied and Numerical Harmonic Analysis. Birkhäuser, Boston (2003)
105. Feichtinger, H.G., Zimmermann, G.: A Banach space of test functions for Gabor analysis. In: Gabor Analysis and Algorithms, pp. 123–170. Birkhäuser, Boston (1998)
106. Fernández, C., Galbis, A.: Compactness of time-frequency localization operators on $L^2(\mathbb{R}^d)$. J. Funct. Anal. **233**(2), 335–350 (2006)
107. Fernández, C., Galbis, A., Martínez, J.: Multilinear Fourier multipliers related to time-frequency localization. J. Math. Anal. Appl. **398**(1), 113–122 (2013)
108. Folland, G.B.: Harmonic analysis in phase space. In: Annals of Mathematical Studies, vol. 122. Princeton University Press, Princeton (1989)
109. Folland, G.B.: Real analysis. Modern Techniques and Their Applications, 2nd edn. Wiley, New York (1999)
110. Forlano, J., Oh, T.: On the uniqueness of the one-dimensional cubic nonlinear Schrödinger equation in almost critical spaces (2018). Preprint
111. Gabor, D.: Theory of communication. J. IEE Lond. **93**, 429–457 (1946)
112. Galperin, Y.V.: On compactness of embeddings of Fourier-Lebesgue spaces into modulation spaces. Int. J. Anal. **2013**, 681573 (2013)
113. Galperin, Y.V., Gröchenig, K.: Uncertainty principles as embeddings of modulation spaces. J. Math. Anal. Appl. **274**(1), 181–202 (2002)
114. Galperin, Y.V., Samarah, S.: Time-frequency analysis on modulation spaces $M_m^{p,q}$, $0 < p, q \leq \infty$. Appl. Comput. Harmon. Anal. **16**(1), 1–18 (2004)
115. García-Cuerva, J., Rubio de Francia, L.: Weighted Norm Inequalities and Related Topics. Elsevier, North-Holland (1985)
116. Gautam, S.Z.: A critical-exponent Balian-Low theorem. Math. Res. Lett. **15**, 471–483 (2008)
117. Gehring, F.W.: The L^p integrability of the partial derivative of a quasiconformal mapping. Acta Math. **130**, 265–277 (1973)
118. Ginibre, J., Velo, G.: Smoothing properties and retarded estimates for some dispersive evolution equations. Commun. Math. Phys. **144**(1), 163–188 (1992)
119. Grafakos, L.: Classical Fourier Analysis, 2nd edn. Graduate Texts in Mathematics, vol. 249. Springer, New York (2008)
120. Grafakos, L.: Modern Fourier Analysis, 2nd edn. Graduate Texts in Mathematics, vol. 250. Springer, New York (2009)

121. Grafakos, L., Li, X.: Uniform bounds for the bilinear Hilbert transforms. I. Ann. Math. **159**(3), 889–993 (2004)
122. Grafakos, L., Torres, R.H.: Multilinear Calderón-Zygmund theory. Adv. Math. **165**, 124–164 (2002)
123. Gröbner, P.: Banachräume glatter Funktionen und zerlegungsmethoden. Ph.D. thesis, University of Vienna, Vienna (1992)
124. Gröchenig, K.: An uncertainty principle related to the Poisson summation formula. Studia Math. **121**(1), 87–104 (1996)
125. Gröchenig, K.: Foundations of Time-Frequency Analysis. Birkhäuser, Boston (2001)
126. Gröchenig, K.: Weight functions in time-frequency analysis. In: Pseudo-Differential Operators, Partial Differential Equations and Time-Frequency Analysis. Fields Institute Communications, vol. 52, pp. 343–366. American Mathematical Society, New York (2007)
127. Gröchenig, K., Heil, C.: Modulation spaces and pseudodifferential operators. Integr. Equ. Oper. Theory **34**(4), 439–457 (1999)
128. Gröchenig, K., Heil, C.: Gabor meets Littlewood-Paley: Gabor expansions in $L^p(\mathbb{R}^d)$. Studia Math. **146**, 115–33 (2001)
129. Gröchenig, K., Heil, C.: Counterexamples for boundedness of pseudodifferential operators. Osaka J. Math. **41**, 681–691 (2004)
130. Gröchenig, K., Leinert, M.: Wiener's lemma for twisted convolution and Gabor frames. J. Am. Math. Soc. **17**(1), 1–18 (2004)
131. Gröchenig, K., Toft, J.: Isomorphism properties of Toeplitz operators and pseudo-differential operators between modulation spaces. J. Anal. Math. **114**, 255–283 (2011)
132. Gröchenig, K., Heil, C., Okoudjou, K.A.: Gabor analysis in weighted amalgam spaces. Sampl. Theory Signal Image Process. **1**(3), 225–259 (2002)
133. Gulisashvili, A., Kon. M.: Exact smoothing properties of Schrödinger semigroups. Am. J. Math. **118**, 1215–1248 (1996)
134. Guo, S.: On the 1d cubic NLS in an almost critical space. Master thesis. University of Bonn, Germany (2014)
135. Guo, S.: On the 1d cubic nonlinear Schrödinger equation in an almost critical space. J. Fourier Anal. Appl. **23**(1), 91–124 (2017)
136. Guo, B., Wang, B.X., Zhao, L.: Isometric decomposition operators, function spaces $E^\lambda_{p,q}$ and applications to nonlinear evolution equations. J. Funct. Anal. **233**, 1–39 (2006)
137. Guo, S., Ren, X., Wang, B.X.: Local well-posedness for the derivative nonlinear Schrödinger equations with L^2 subcritical data (2016). arXiv:1608.03136
138. Guo, W., Fan, D., Wu, H., Zhao, G.: Sharp weighted convolution inequalities and some applications. Studia Math. **241**(3), 201–239 (2018)
139. Han, J., Wang, B.X.: α-Modulation spaces (I) scaling, embedding and algebraic properties. J. Math. Soc. Jpn. **66**(4), 1315–1373 (2014)
140. Han, J., Wang, B.X.: Global well-posedness for NLS with a class of H^s-supercritical data (2019). arXiv:1901.08868
141. Han, L., Huang, C., Wang, B.X.: Global well-posedness and scattering for the derivative nonlinear Schrödinger equation with small rough data. Ann. I.H. Poincaré **26**, 2253–2281 (2009)
142. Hao, C., Huo, Z., Guo, Z., Wang, B.X.: Harmonic Analysis Method for Nonlinear Evolution Equations. I. World Scientific, Hackensack (2011)
143. Heil, C., Powel, A.: Gabor Schauder bases and the Balian-Low theorem. J. Math. Phys. **47**(11), 113506, 21 pp. (2006)
144. Heil, C., Tinaztepe, R.: Modulation spaces, BMO, and the Balian-Low theorem. Sampl. Theory Signal Image Process. **11**(1), 25–41 (2012)
145. Heil, C., Walnut, D.: Continuous and discrete wavelet transforms. SIAM Rev. **31**(4), 628–666 (1989)
146. Heil, C., Ramanathan, J., Topiwala, P.: Linear independence of time-frequency translates. Proc. Am. Math. Soc. **124**(9), 2787–2795 (1996)

147. Heil, C., Ramanathan, J., Topiwala, P.: Singular values of compact pseudodifferential operators. J. Funct. Anal. **150**(2), 426–452 (1997)

148. Hernández, E., Weiss, G.: A First Course on Wavelets. CRC, Boca Raton (1996)

149. Hogan, J.A., Lakey, J.D: Embeddings and uncertainty principles for generalized modulation spaces. In: Modern Sampling Theory. Application Numerical Harmon. Anal., pp. 73–105. Birkhäuser, Boston (2001)

150. Hörmander, L.: Estimates for translation invariant operators in L^p spaces. Acta Math. **104**, 93–139 (1960)

151. Hörmander, L.: Pseudo-differential operators. Commun. Pure Appl. Math. **18**, 501–517 (1965)

152. Hörmander, L.: Pseudo-differential operators and non-elliptic boundary problems. Ann. Math. **83**, 129–209 (1966)

153. Hörmander, L.: Pseudo-differential operators and hypoelliptic equations. In: Singular Integrals. Proceedings of Symposia in Pure Mathematics, Chicago, Ill., 1966, vol. X, pp. 138–183. American Mathematical Society, Providence (1967)

154. Huang, C., Wang, B.X.: Frequency-uniform decomposition method for the generalized BO, KdV and NLS equations. J. Differ. Equ. **239**, 213–250 (2007)

155. Hudzik, H., Wang, B.X.: The global Cauchy problem for the NLS and NLKG with small rough data. J. Differ. Equ. **232**(1), 36–73 (2007)

156. Hytönen, T.: The sharp weighted bound for general Calderón-Zygmund operators. Ann. Math. **175**(3), 1473–1506 (2012)

157. Jakobsen, M.S.: On a (no longer) new Segal algebra—a review of the Feichtinger algebra. J. Fourier Anal. Appl. **24**(6), 1579–1660 (2018)

158. Kahane, J.P.: Series de Fourier absolument convergentes. Springer, Berlin (1970)

159. Kato, T., Ponce, G.: Commutator estimates and the Euler and Navier-Stokes equations. Comm. Pure Appl. Math **41**, 891–907 (1988)

160. Kato, K., Kobayashi, M., Ito, S.: Remark on wave front sets of solutions to Schrödinger equation of a free particle and a harmonic oscillator. SUT J. Math. **47**, 175–183 (2011)

161. Kato, K., Kobayashi, M., Ito, S.: Representation of Schrödinger operator of a free particle via short time Fourier transform and its applications. Tohoku Math. J. **64**, 223–231 (2012)

162. Kato, K., Kobayashi, M., Ito, S.: Remarks on Wiener amalgam space type estimates for the Schrödinger equation. In: RIMS Kôkyûroku Bessatsu, vol. B33, pp. 41–48. Research Institute for Mathematical Sciences, Kyoto (2012)

163. Kato, K., Kobayashi, M., Ito, S.: Estimates on modulation spaces for Schrödinger evolution operators with quadratic and sub-quadratic potentials. J. Funct. Anal. **266**(2), 733–753 (2014)

164. Kato, T., Sugimoto, M., Tomita, N.: Nonlinear operations on a class of modulation spaces (2018). arXiv: 1801.06803v1

165. Keel, M., Tao, T.: Endpoint Strichartz estimates. Am. J. Math. **120**(5), 955–980 (1998)

166. Kobayashi, M.: Modulation spaces $M^{p,q}$ for $0 < p, q \leq \infty$. J. Funct. Spaces Appl. **4**(3), 329–341 (2006)

167. Kobayashi, M.: Dual of modulation spaces. J. Funct. Spaces Appl. **5**(1), 1–8 (2007)

168. Kobayashi, M., Sato, E.: Operating functions on modulation and Wiener amalgam spaces. Nagoya Math. J. **230**, 72–82 (2018)

169. Kobayashi, M., Sawano, Y.: Molecular decomposition of the modulation spaces. Osaka J. Math. **47**, 1029–1053 (2010)

170. Kobayashi, M., Sugimoto, M.: The inclusion relation between Sobolev and modulation spaces. J. Funct. Anal. **260**(11), 3189–3208 (2011)

171. Kobayashi, M., Miyachi, A., Tomita, N.: Embedding relations between local Hardy and modulation spaces. Studia Math. **192**, 79–96 (2009)

172. Kohn, J.J., Nirenberg, L.: An algebra of pseudo-differential operators. Commun. Pure Appl. Math. **18**, 269–305 (1965)

173. Krishtal, I., Okoudjou, K.A.: Invertibility of the Gabor frame operator on the Wiener amalgam space. J. Approx. Theory **153**(2), 212–214 (2008)

174. Lacey, M., Thiele, C.: L^p bounds for the bilinear Hilbert transform, $2 < p < \infty$. Ann. Math. **146**, 693–724 (1997)

175. Lacey, M., Thiele, C.: On Calderón's conjecture. Ann. Math. **149**, 475–496 (1999)

176. Lebedev, V., Olevskiĭ, A.: C^1 changes of variable: Beurling-Helson type theorem and Hörmander conjecture on Fourier multipliers. Geom. Funct. Anal. **4**(2), 213–235 (1994)

177. Lieb, E.H.: Integral bounds for radar ambiguity functions and Wigner distributions. J. Math. Phys. **31**(3), 594–599 (1990)

178. Lieb, E.H., Michael, L.: Analysis, 2nd edn. Graduate Studies in Mathematics, vol. 14. American Mathematical Society, New York (2001)

179. Lührmann, J., Mendelson, D.: Random data Cauchy theory for nonlinear wave equations of power-type on $\mathbb{R}^3$. Commun. Partial Differ. Equ. **39**(12), 2262–2283 (2014)

180. Meyer, Y.: Remarques sur un théorème de J.-M. Bony. In: Proceedings of the Seminar on Harmonic Analysis (Pisa, 1980). Rendiconti del Circolo Matematico di Palermo, vol. 2(1), pp. 1–20 (1981)

181. Michalowski, N., Rule, D., Staubach, W.: Multilinear pseudodifferential operators beyond Calderón-Zygmund theory. J. Math. Anal. Appl. **414**(1), 149–165 (2014)

182. Miyachi, A.: On some Fourier multipliers for $H^p(\mathbb{R}^n)$. J. Fac. Sci. Univ. Tokyo Sec. IA Math. **27**, 157–179 (1980)

183. Miyachi, A., Tomita, N.: Calderón-Vaillancourt type theorem for bilinear pseudo-differential operators. Indiana Univ. Math. J. **62**(4), 1165–1201 (2013)

184. Miyachi, A., Nicola, F., Rivetti, S., Tabacco, A., Tomita, N.: Estimates for unimodular Fourier multipliers on modulation spaces. Proc. Am. Math. Soc. **137**(11), 3869–3883 (2009)

185. Molahajloo, S., Okoudjou, K.A., Pfander, G.E.: Boundedness of multilinear pseudo-differential operators on modulation spaces. J. Fourier Anal. Appl. **22**(6), 1381–1415 (2016)

186. Muckenhoupt, B.: Weighted norm inequalities for the Hardy maximal function. Trans. Am. Math. Soc. **165**, 207–226 (1972)

187. Muckenhoupt, B., Wheeden, R.L.: Weighted norm inequalities for singular and fractional integrals. Trans. Am. Math. Soc **161**, 249–258 (1971)

188. Muckenhoupt, B., Wheeden, R.L.: Weighted bounded mean oscillation and the Hilbert transform. Studia Math. **54**, 221–237 (1976)

189. Oh, T.: Nonlinear Schrödinger equations/Dispersive equations. In: MIGSAA advanced Ph.D. course, University of Edinburgh (2018). https://www.maths.ed.ac.uk/~toh/2018NSE.html

190. Oh, T., Pocovnicu, O.: Probabilistic global well-posedness of the energy-critical defocusing quintic nonlinear wave equation on $\mathbb{R}^3$. J. Math. Pures Appl. **105**(3), 342–366 (2016)

191. Oh, T., Wang, Y.: On global well-posedness of the modified KDV equation in modulation spaces (2018). arXiv:1811.04606v1

192. Oh, T., Wang, Y.: Global well-posedness of the one-dimensional cubic nonlinear Schrödinger equation in almost critical spaces (2018). arXiv:1806.08761

193. Okoudjou, K.A.: Embeddings of some classical Banach spaces into modulation spaces. Proc. Am. Math. Soc. **132**, 1639–1647 (2004)

194. Okoudjou, K.A.: A Beurling-Helson type theorem for modulation spaces. J. Funct. Spaces Appl. **7**(1), 33–41 (2009)

195. Okoudjou, K.A.: An invitation to Gabor analysis. Notices Am. Math. Soc. **66**(6), 808–819 (2019)

196. Pitt, H.R.: Theorems on Fourier series and power series. Duke Math. J. **3**, 747–755 (1937)

197. Pocovnicu, O.: Almost sure global well-posedness for the energy-critical defocusing nonlinear wave equation on $\mathbb{R}^d$, $d = 4$ and 5. J. Eur. Math. Soc. **19**(8), 2521–2575 (2017)

198. Rauhut, H.: Coorbit space theory for quasi-Banach spaces. Studia Math. **180**(3), 237–253 (2007)

199. Ron, A., Shen, Z.: Weyl-Heisenberg frames and Riesz bases in $L_2(\mathbb{R}^d)$. Duke Math. J. **89**(2), 237–282 (1997)

200. Rudin, W.: Functional Analysis. McGraw-Hill Series in Higher Mathematics. McGraw-Hill Book, New York (1973)

201. Runst, T.: Paradifferential operators in spaces of Triebel-Lizorkin and Besov type. Z. Anal. Anwendungen **4**, 557–573 (1985)
202. Runst, T., Sickel, W.: Sobolev spaces of fractional order, Nemytskij operators, and nonlinear partial differential equations. In: de Gruyter Series in Nonlinear Analysis and Applications, vol. 3, Walter de Gruyter, Berlin (1996)
203. Ruzhansky, M., Sugimoto, M., Toft, J., Tomita, N.: Changes of variables in modulation and Wiener amalgam spaces. Math. Nachr. **284**(16), 2078–2092 (2011)
204. Ruzhansky, M., Sugimoto, M., Wang, B.X.: Modulation spaces and nonlinear evolution equations. In: Evolution Equations of Hyperbolic and Schrödinger Type. Progress in Mathematics, vol. 301, pp. 267–283. Birkhäuser/Springer, Basel (2012)
205. Self, W.M.: Some consequences of the Beurling-Helson theorem. Rocky Mountain J. Math. **6**, 177–180 (1976)
206. Sjöstrand, J.: An algebra of pseudodifferential operators. Math. Res. Lett. **1**(2), 185–192 (1994)
207. Song, C.: Unimodular Fourier multipliers with a time parameter on modulation spaces. J. Inequal. Appl. **43**, 15 (2014)
208. Stein, E.M.: Harmonic Analysis: Real-Variable Methods, Orthogonality, and Oscillatory Integrals. Princeton University Press, Princeton (1993)
209. Stein, E.M., Shakarchi, R.: Functional Analysis: Introduction to Further Topics in Analysis. Princeton University Press, Princeton (2011)
210. Strichartz, R.S.: Restrictions of Fourier transforms to quadratic surfaces and decay of solutions of wave equations. Duke Math. J. **44**(3), 705–714 (1977)
211. Sugimoto, M., Tomita, N.: The dilation property of modulation spaces and their inclusion relation with Besov spaces. J. Funct. Anal. **248**(1), 79–106 (2007)
212. Sugimoto, M., Tomita, N.: A counterexample for boundedness of pseudo-differential operators on modulation spaces. Proc. Am. Math. Soc. **136**(5), 1681–1690 (2008)
213. Sugimoto, M., Tomita, N.: Boundedness properties of pseudo-differential and Calderón-Zygmund operators on modulation spaces. J. Fourier Anal. Appl. **14**(1), 124–143 (2008)
214. Sugimoto, M., Tomita, N., Wang, B.X.: Remarks on nonlinear operations on modulation spaces. Integral Transforms Spec. Funct. **22**(4–5), 351–358 (2011)
215. Tao, T.: Global regularity of wave maps II. Small energy in two dimensions. Commun. Math. Phys. **224**(2), 443–544 (2001)
216. Tao, T.: Nonlinear dispersive equations. Local and global analysis. In: CBMS Regional Conference Series in Mathematics, vol. 106. American Mathematical Society, New York (2006)
217. Taylor, M.E.: Pseudodifferential Operators and Nonlinear PDEs. Birkhäuser, Boston (1991)
218. Taylor, M.E.: Tools for PDE: Pseudodifferential Operators, Paradifferential Operators, and Layer Potentials. Mathematical Surveys and Monographs, vol. 81. American Mathematical Society, New York (2000)
219. Toft, J.: Convolutions and embeddings for weighted modulation spaces. In: Advances in Pseudodifferential Operators, pp. 165–186. Birkhäuser, Basel (2004)
220. Toft, J.: Continuity properties for modulation spaces, with applications to pseudo-differential calculus, I. J. Funct. Anal. **207**, 399–429 (2004)
221. Toft, J.: Continuity properties for modulation spaces, with applications to pseudo-differential calculus, II. Ann. Global Anal. Geom. **26**(1), 73–106 (2004)
222. Toft, J.: Pseudo-differential operators with smooth symbols on modulation spaces. Cubo **11**(4), 87–107 (2009)
223. Toft, J., Wahlberg, P.: Embeddings of α-modulation spaces. Pliska Stud. Math. Bulgar. **21**, 25–46 (2012)
224. Tomita, N.: Unimodular Fourier multipliers on modulation spaces $M^{p,q}$ for $0 < p < 1$. In: Harmonic Analysis and Nonlinear Partial Differential Equations. RIMS Kôkyûroku Bessatsu, vol. B18, pp. 125–131. Research Institute for Mathematical Sciences, Kyoto (2010)
225. Triebel, H.: Fourier Analysis and Function Spaces (Selected Topics). Teubner, Leipzig (1977)
226. Triebel, H.: Spaces of Besov-Hardy-Sobolev Type. B. G. Teubner, Leipzig (1978)

227. Triebel, H.: Theory of Function Spaces. Birkhäuser, Basel (1983)
228. Triebel, H.: Modulation spaces on the Euclidean n-space. Z. Anal. Anwendungen **2**(5), 443–457 (1983)
229. Triebel, H.: Theory of Function Spaces, II. Birkhäuser, Basel (1992)
230. Walnut, D.F.: Continuity properties of the Gabor frame operator. J. Math. Anal. Appl. **165**(2), 479–504 (1992)
231. Wong, M.W.: Wavelet transforms and localization operators. In: Operator Theory: Advances and Applications, vol. 136. Birkhäuser, Basel (2002)
232. Yajima, K.: Existence of solutions for Schrödinger evolution equations. Commun. Math. Phys. **110**(3), 415–426 (1987)

Index

© Springer Science+Business Media, LLC, part of Springer Nature 2020
Á. Bényi, K. A. Okoudjou, *Modulation Spaces*, Applied and Numerical
Harmonic Analysis, https://doi.org/10.1007/978-1-0716-0332-1

Applied and Numerical Harmonic Analysis (98 volumes)

1. A. I. Saichev and W. A. Woyczyñski: *Distributions in the Physical and Engineering Sciences* (ISBN: 978-0-8176-3924-2)
2. C. E. D'Attellis and E. M. Fernandez-Berdaguer: *Wavelet Theory and Harmonic Analysis in Applied Sciences* (ISBN: 978-0-8176-3953-2)
3. H. G. Feichtinger and T. Strohmer: *Gabor Analysis and Algorithms* (ISBN: 978-0-8176-3959-4)
4. R. Tolimieri and M. An: *Time-Frequency Representations* (ISBN: 978-0-8176-3918-1)
5. T. M. Peters and J. C. Williams: *The Fourier Transform in Biomedical Engineering* (ISBN: 978-0-8176-3941-9)
6. G. T. Herman: *Geometry of Digital Spaces* (ISBN: 978-0-8176-3897-9)
7. A. Teolis: *Computational Signal Processing with Wavelets* (ISBN: 978-0-8176-3909-9)
8. J. Ramanathan: *Methods of Applied Fourier Analysis* (ISBN: 978-0-8176-3963-1)
9. J. M. Cooper: *Introduction to Partial Differential Equations with MATLAB* (ISBN: 978-0-8176-3967-9)
10. Procházka, N. G. Kingsbury, P. J. Payner, and J. Uhlir: *Signal Analysis and Prediction* (ISBN: 978-0-8176-4042-2)
11. W. Bray and C. Stanojevic: *Analysis of Divergence* (ISBN: 978-1-4612-7467-4)
12. G. T. Herman and A. Kuba: *Discrete Tomography* (ISBN: 978-0-8176-4101-6)
13. K. Gröchenig: *Foundations of Time-Frequency Analysis* (ISBN: 978-0-8176-4022-4)
14. L. Debnath: *Wavelet Transforms and Time-Frequency Signal Analysis* (ISBN: 978-0-8176-4104-7)
15. J. J. Benedetto and P. J. S. G. Ferreira: *Modern Sampling Theory* (ISBN: 978-0-8176-4023-1)
16. D. F. Walnut: *An Introduction to Wavelet Analysis* (ISBN: 978-0-8176-3962-4)

© Springer Science+Business Media, LLC, part of Springer Nature 2020
Á. Bényi, K. A. Okoudjou, *Modulation Spaces*, Applied and Numerical Harmonic Analysis, https://doi.org/10.1007/978-1-0716-0332-1

17. A. Abbate, C. DeCusatis, and P. K. Das: *Wavelets and Subbands* (ISBN: 978-0-8176-4136-8)
18. O. Bratteli, P. Jorgensen, and B. Treadway: *Wavelets Through a Looking Glass* (ISBN: 978-0-8176-4280-80
19. H. G. Feichtinger and T. Strohmer: *Advances in Gabor Analysis* (ISBN: 978-0-8176-4239-6)
20. O. Christensen: *An Introduction to Frames and Riesz Bases* (ISBN: 978-0-8176-4295-2)
21. L. Debnath: *Wavelets and Signal Processing* (ISBN: 978-0-8176-4235-8)
22. G. Bi and Y. Zeng: *Transforms and Fast Algorithms for Signal Analysis and Representations* (ISBN: 978-0-8176-4279-2)
23. J. H. Davis: *Methods of Applied Mathematics with a MATLAB Overview* (ISBN: 978-0-8176-4331-7)
24. J. J. Benedetto and A. I. Zayed: *Sampling, Wavelets, and Tomography* (ISBN: 978-0-8176-4304-1)
25. E. Prestini: *The Evolution of Applied Harmonic Analysis* (ISBN: 978-0-8176-4125-2)
26. L. Brandolini, L. Colzani, A. Iosevich, and G. Travaglini: *Fourier Analysis and Convexity* (ISBN: 978-0-8176-3263-2)
27. W. Freeden and V. Michel: *Multiscale Potential Theory* (ISBN: 978-0-8176-4105-4)
28. O. Christensen and K. L. Christensen: *Approximation Theory* (ISBN: 978-0-8176-3600-5)
29. O. Calin and D.-C. Chang: *Geometric Mechanics on Riemannian Manifolds* (ISBN: 978-0-8176-4354-6)
30. J. A. Hogan: *Time?Frequency and Time?Scale Methods* (ISBN: 978-0-8176-4276-1)
31. C. Heil: *Harmonic Analysis and Applications* (ISBN: 978-0-8176-3778-1)
32. K. Borre, D. M. Akos, N. Bertelsen, P. Rinder, and S. H. Jensen: *A Software-Defined GPS and Galileo Receiver* (ISBN: 978-0-8176-4390-4)
33. T. Qian, M. I. Vai, and Y. Xu: *Wavelet Analysis and Applications* (ISBN: 978-3-7643-7777-9)
34. G. T. Herman and A. Kuba: *Advances in Discrete Tomography and Its Applications* (ISBN: 978-0-8176-3614-2)
35. M. C. Fu, R. A. Jarrow, J.-Y. Yen, and R. J. Elliott: *Advances in Mathematical Finance* (ISBN: 978-0-8176-4544-1)
36. O. Christensen: *Frames and Bases* (ISBN: 978-0-8176-4677-6)
37. P. E. T. Jorgensen, J. D. Merrill, and J. A. Packer: *Representations, Wavelets, and Frames* (ISBN: 978-0-8176-4682-0)
38. M. An, A. K. Brodzik, and R. Tolimieri: *Ideal Sequence Design in Time-Frequency Space* (ISBN: 978-0-8176-4737-7)
39. S. G. Krantz: *Explorations in Harmonic Analysis* (ISBN: 978-0-8176-4668-4)
40. B. Luong: *Fourier Analysis on Finite Abelian Groups* (ISBN: 978-0-8176-4915-9)

41. G. S. Chirikjian: *Stochastic Models, Information Theory, and Lie Groups, Volume 1* (ISBN: 978-0-8176-4802-2)
42. C. Cabrelli and J. L. Torrea: *Recent Developments in Real and Harmonic Analysis* (ISBN: 978-0-8176-4531-1)
43. M. V. Wickerhauser: *Mathematics for Multimedia* (ISBN: 978-0-8176-4879-4)
44. B. Forster, P. Massopust, O. Christensen, K. Gröchenig, D. Labate, P. Vandergheynst, G. Weiss, and Y. Wiaux: *Four Short Courses on Harmonic Analysis* (ISBN: 978-0-8176-4890-9)
45. O. Christensen: *Functions, Spaces, and Expansions* (ISBN: 978-0-8176-4979-1)
46. J. Barral and S. Seuret: *Recent Developments in Fractals and Related Fields* (ISBN: 978-0-8176-4887-9)
47. O. Calin, D.-C. Chang, and K. Furutani, and C. Iwasaki: *Heat Kernels for Elliptic and Sub-elliptic Operators* (ISBN: 978-0-8176-4994-4)
48. C. Heil: *A Basis Theory Primer* (ISBN: 978-0-8176-4686-8)
49. J. R. Klauder: *A Modern Approach to Functional Integration* (ISBN: 978-0-8176-4790-2)
50. J. Cohen and A. I. Zayed: *Wavelets and Multiscale Analysis* (ISBN: 978-0-8176-8094-7)
51. D. Joyner and J.-L. Kim: *Selected Unsolved Problems in Coding Theory* (ISBN: 978-0-8176-8255-2)
52. G. S. Chirikjian: *Stochastic Models, Information Theory, and Lie Groups, Volume 2* (ISBN: 978-0-8176-4943-2)
53. J. A. Hogan and J. D. Lakey: *Duration and Bandwidth Limiting* (ISBN: 978-0-8176-8306-1)
54. G. Kutyniok and D. Labate: *Shearlets* (ISBN: 978-0-8176-8315-3)
55. P. G. Casazza and P. Kutyniok: *Finite Frames* (ISBN: 978-0-8176-8372-6)
56. V. Michel: *Lectures on Constructive Approximation* (ISBN : 978-0-8176-8402-0)
57. D. Mitrea, I. Mitrea, M. Mitrea, and S. Monniaux: *Groupoid Metrization Theory* (ISBN: 978-0-8176-8396-2)
58. T. D. Andrews, R. Balan, J. J. Benedetto, W. Czaja, and K. A. Okoudjou: *Excursions in Harmonic Analysis, Volume 1* (ISBN: 978-0-8176-8375-7)
59. T. D. Andrews, R. Balan, J. J. Benedetto, W. Czaja, and K. A. Okoudjou: *Excursions in Harmonic Analysis, Volume 2* (ISBN: 978-0-8176-8378-8)
60. D. V. Cruz-Uribe and A. Fiorenza: *Variable Lebesgue Spaces* (ISBN: 978-3-0348-0547-6)
61. W. Freeden and M. Gutting: *Special Functions of Mathematical (Geo-)Physics* (ISBN: 978-3-0348-0562-9)
62. A. I. Saichev and W. A. Woyczyñski: *Distributions in the Physical and Engineering Sciences, Volume 2: Linear and Nonlinear Dynamics of Continuous Media* (ISBN: 978-0-8176-3942-6)
63. S. Foucart and H. Rauhut: *A Mathematical Introduction to Compressive Sensing* (ISBN: 978-0-8176-4947-0)

64. G. T. Herman and J. Frank: *Computational Methods for Three-Dimensional Microscopy Reconstruction* (ISBN: 978-1-4614-9520-8)
65. A. Paprotny and M. Thess: *Realtime Data Mining: Self-Learning Techniques for Recommendation Engines* (ISBN: 978-3-319-01320-6)
66. A. I. Zayed and G. Schmeisser: *New Perspectives on Approximation and Sampling Theory: Festschrift in Honor of Paul Butzer's 85th Birthday* (ISBN: 978-3-319-08800-6)
67. R. Balan, M. Begue, J. Benedetto, W. Czaja, and K. A. Okoudjou: *Excursions in Harmonic Analysis, Volume 3* (ISBN: 978-3-319-13229-7)
68. H. Boche, R. Calderbank, G. Kutyniok, and J. Vybiral: *Compressed Sensing and its Applications* (ISBN: 978-3-319-16041-2)
69. S. Dahlke, F. De Mari, P. Grohs, and D. Labate: *Harmonic and Applied Analysis: From Groups to Signals* (ISBN: 978-3-319-18862-1)
70. A. Aldroubi: *New Trends in Applied Harmonic Analysis* (ISBN: 978-3-319-27871-1)
71. M. Ruzhansky: *Methods of Fourier Analysis and Approximation Theory* (ISBN: 978-3-319-27465-2)
72. G. Pfander: *Sampling Theory, a Renaissance* (ISBN: 978-3-319-19748-7)
73. R. Balan, M. Begue, J. Benedetto, W. Czaja, and K. A. Okoudjou: *Excursions in Harmonic Analysis, Volume 4* (ISBN: 978-3-319-20187-0)
74. O. Christensen: *An Introduction to Frames and Riesz Bases, Second Edition* (ISBN: 978-3-319-25611-5)
75. E. Prestini: *The Evolution of Applied Harmonic Analysis: Models of the Real World, Second Edition* (ISBN: 978-1-4899-7987-2)
76. J. H. Davis: *Methods of Applied Mathematics with a Software Overview, Second Edition* (ISBN: 978-3-319-43369-1)
77. M. Gilman, E. M. Smith, and S. M. Tsynkov: *Transionospheric Synthetic Aperture Imaging* (ISBN: 978-3-319-52125-1)
78. S. Chanillo, B. Franchi, G. Lu, C. Perez, and E. T. Sawyer: *Harmonic Analysis, Partial Differential Equations and Applications* (ISBN: 978-3-319-52741-3)
79. R. Balan, J. Benedetto, W. Czaja, M. Dellatorre, and K. A. Okoudjou: *Excursions in Harmonic Analysis, Volume 5* (ISBN: 978-3-319-54710-7)
80. I. Pesenson, Q. T. Le Gia, A. Mayeli, H. Mhaskar, and D. X. Zhou: *Frames and Other Bases in Abstract and Function Spaces: Novel Methods in Harmonic Analysis, Volume 1* (ISBN: 978-3-319-55549-2)
81. I. Pesenson, Q. T. Le Gia, A. Mayeli, H. Mhaskar, and D. X. Zhou: *Recent Applications of Harmonic Analysis to Function Spaces, Differential Equations, and Data Science: Novel Methods in Harmonic Analysis, Volume 2* (ISBN: 978-3-319-55555-3)
82. F. Weisz: *Convergence and Summability of Fourier Transforms and Hardy Spaces* (ISBN: 978-3-319-56813-3)
83. C. Heil: *Metrics, Norms, Inner Products, and Operator Theory* (ISBN: 978-3-319-65321-1)
84. S. Waldron: *An Introduction to Finite Tight Frames: Theory and Applications.* (ISBN: 978-0-8176-4814-5)

85. D. Joyner and C. G. Melles: *Adventures in Graph Theory: A Bridge to Advanced Mathematics.* (ISBN: 978-3-319-68381-2)
86. B. Han: *Framelets and Wavelets: Algorithms, Analysis, and Applications* (ISBN: 978-3-319-68529-8)
87. H. Boche, G. Caire, R. Calderbank, M. März, G. Kutyniok, and R. Mathar: *Compressed Sensing and Its Applications* (ISBN: 978-3-319-69801-4)
88. A. I. Saichev and W. A. Woyczyñski: *Distributions in the Physical and Engineering Sciences, Volume 3: Random and Fractal Signals and Fields* (ISBN: 978-3-319-92584-4)
89. G. Plonka, D. Potts, G. Steidl, and M. Tasche: *Numerical Fourier Analysis* (978-3-030-04305-6)
90. K. Bredies and D. Lorenz: *Mathematical Image Processing* (ISBN: 978-3-030-01457-5)
91. H. G. Feichtinger, P. Boggiatto, E. Cordero, M. de Gosson, F. Nicola, A. Oliaro, and A. Tabacco: *Landscapes of Time-Frequency Analysis* (ISBN: 978-3-030-05209-6)
92. E. Liflyand: *Functions of Bounded Variation and Their Fourier Transforms* (978-3-030-04428-2)
93. R. Campos: *The XFT Quadrature in Discrete Fourier Analysis* (978-3-030-13422-8)
94. M. Abell, E. Iacob, A. Stokolos, S. Taylor, S. Tikhonov, J. Zhu: *Topics in Classical and Modern Analysis: In Memory of Yingkang Hu* (978-3-030-12276-8)
95. H. Boche, G. Caire, R. Calderbank, G. Kutyniok, R. Mathar, P. Petersen: *Compressed Sensing and its Applications: Third International MATHEON Conference 2017* (978-3-319-73073-8)
96. A. Aldroubi, C. Cabrelli, S. Jaffard, U. Molter: *New Trends in Applied Harmonic Analysis, Volume II: Harmonic Analysis, Geometric Measure Theory, and Applications* (978-3-030-32352-3)
97. S. Dos Santos, M. Maslouhi, K. Okoudjou: *Recent Advances in Mathematics and Technology: Proceedings of the First International Conference on Technology, Engineering, and Mathematics, Kenitra, Morocco, March 26-27, 2018* (978-3-030-35201-1)
98. Á. Bényi, K. Okoudjou: *Modulation Spaces: With Applications to Pseudodifferential Operators and Nonlinear Schrödinger Equations* (978-1-0716-0330-7)

For an up-to-date list of ANHA titles, please visit http://www.springer.com/series/4968